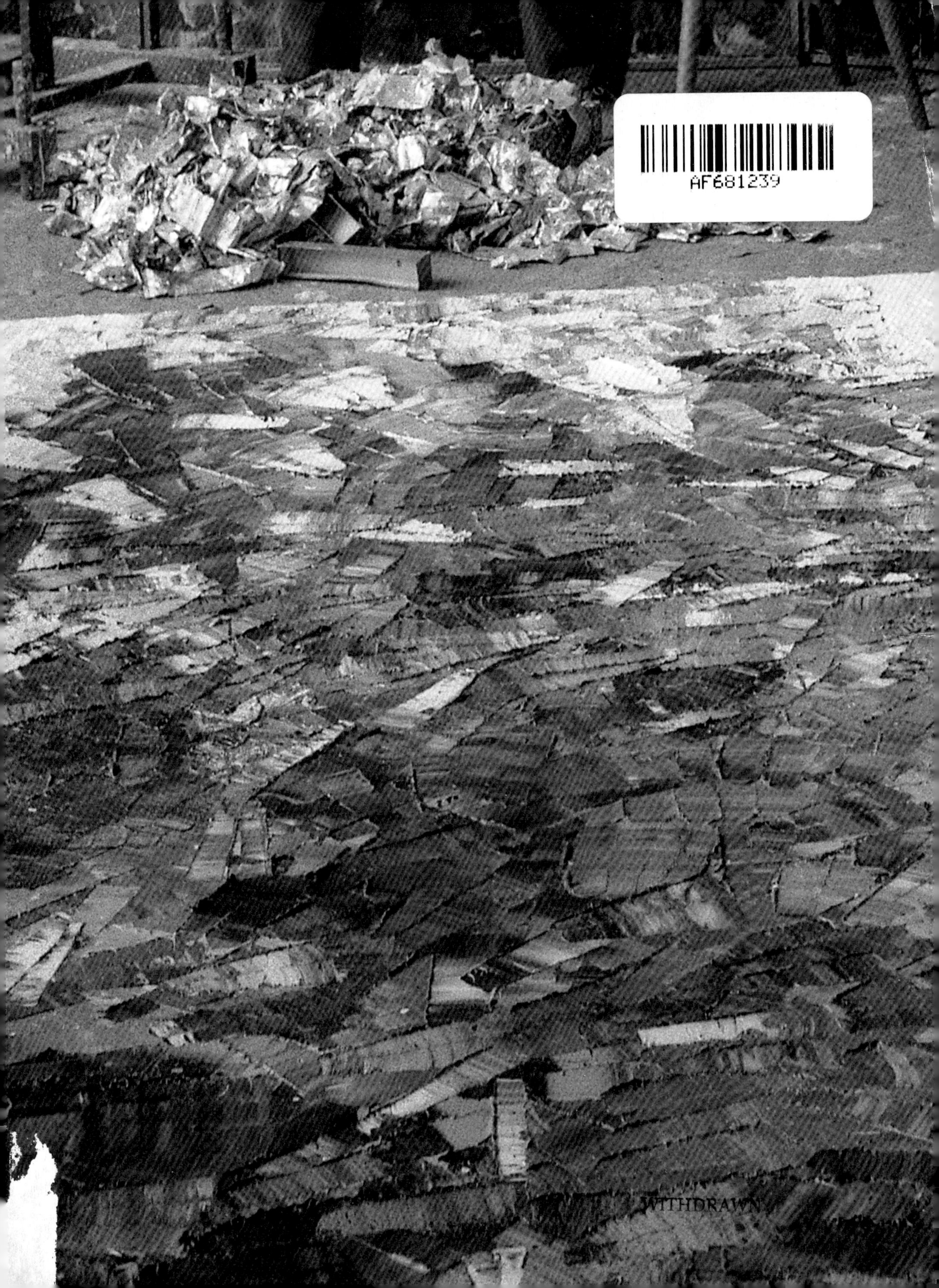

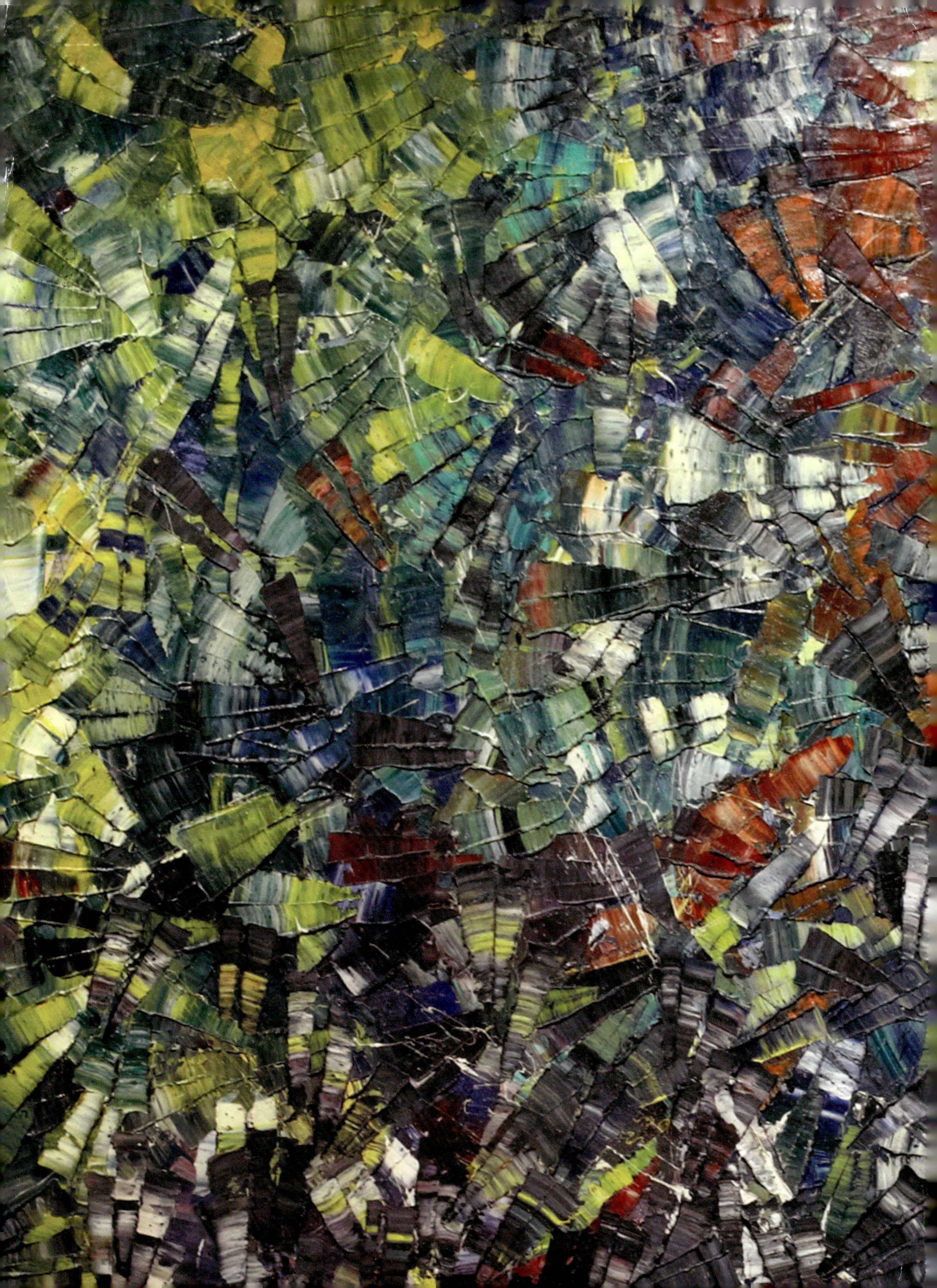

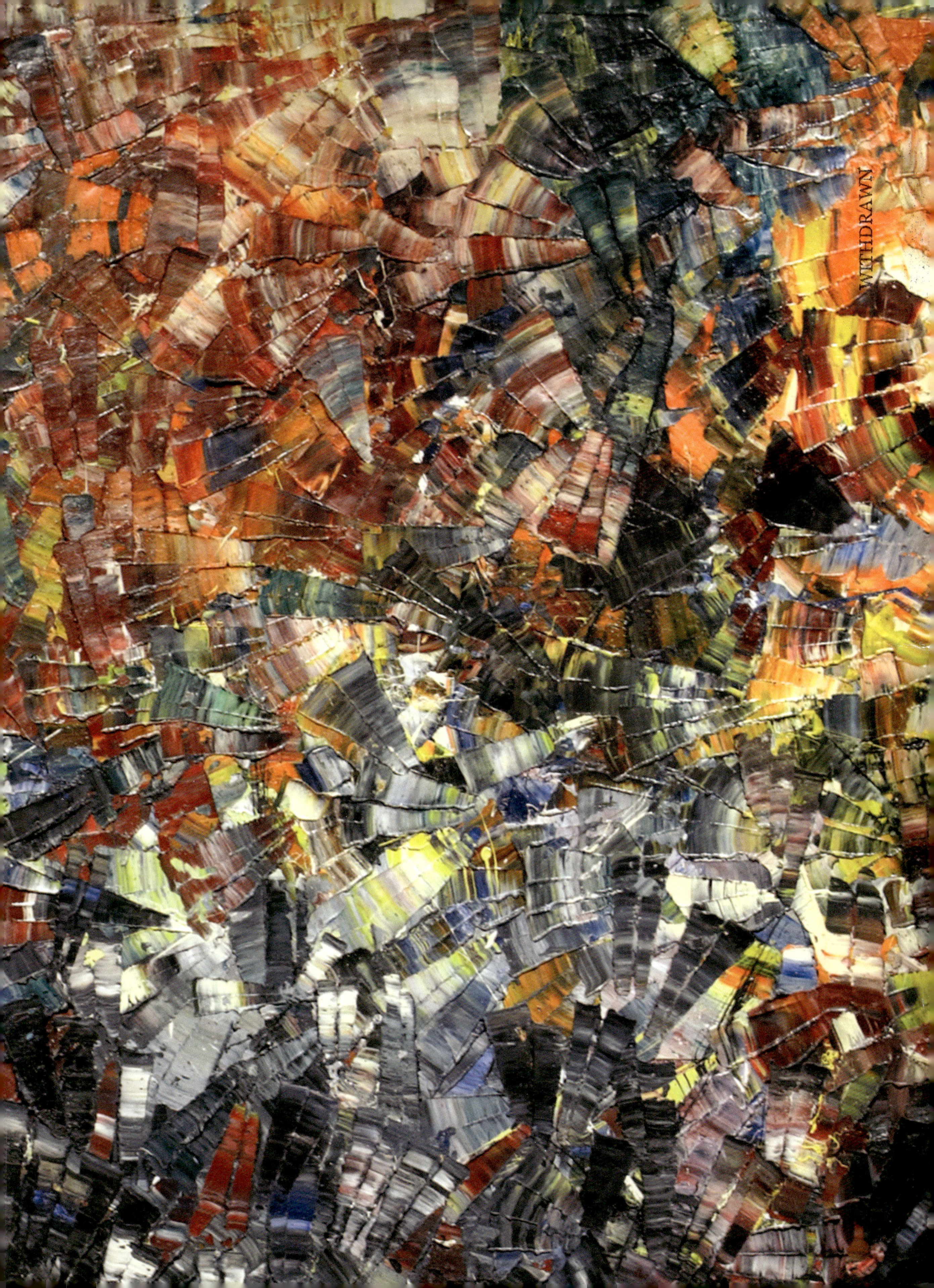

MARIE-CLAUDE CORBEIL
KATE HELWIG
JENNIFER POULIN

JEAN PAUL RIOPELLE

THE ARTIST'S MATERIALS

With an essay by
Stéphane Aquin

Preface by
Yseult Riopelle

The Getty Conservation Institute
Los Angeles

The Getty Conservation Institute

The Getty Conservation Institute works internationally to advance conservation practice in the visual arts—broadly interpreted to include objects, collections, architecture, and sites. The Institute serves the conservation community through scientific research, education and training, model field projects, and the dissemination of the results of both its own work and the work of others in the field. In all its endeavors, the GCI focuses on the creation and delivery of knowledge that will benefit the professionals and organizations responsible for the conservation of the world's cultural heritage.

Published by the Getty Conservation Institute
Getty Publications
1200 Getty Center Drive, Suite 500
Los Angeles, California 90049-1682
www.gettypublications.org

Tevvy Ball, *Project Editor*
Katya Rice, *Manuscript Editor*
Jeffrey Cohen, *Designer*
Suzanne Watson, *Production Coordinator*

Printed in China

Library of Congress Cataloging-in-Publication Data
Corbeil, Marie-Claude, 1960–
Jean Paul Riopelle : the artist's materials / Marie-Claude Corbeil, Kate Helwig, and Jennifer Poulin ; with an essay by Stephane Aquin and a preface by Yseult Riopelle.
p. cm.
Includes bibliographical references and index.
ISBN 978-1-60606-086-5 (pbk.)
1. Riopelle, Jean Paul. 2. Mixed media painting—Technique. 3. Paint materials—Analysis. 4. Artists' materials—Analysis. I. Riopelle, Jean Paul. II. Helwig, Kate. III. Poulin, Jennifer, 1982– IV. Title.
ND249.R56C67 2011
709.2—dc22

2011016447

FRONT COVER: Jean Paul Riopelle, *Rencontre*, 1956 (detail). See fig. 5.
PAGE i: Interior of Riopelle's studio in the Paris suburb of Vanves, 1956 (detail). See fig. 2.
PAGE ii–iii: Jean Paul Riopelle, *Cheuvreuse II*, 1953 (detail). See fig. 4.
PAGE xiv: Jean Paul Riopelle, *Sans titre*, 1953 (detail). See fig. 12.
BACK COVER AND PAGE x: Riopelle in his studio in Paris, 1952 (detail).
Photo by John Craven. © John Craven. Courtesy Galerie Berthet-Aittouarès, Paris.

Contents

Foreword

This volume is the second in the Getty Conservation Institute's Artist's Materials series, which focuses on the study and conservation of materials used by modern artists. The series features highly celebrated figures such as Willem de Kooning, as well as others who may not be household names. Although Jean Paul Riopelle is perhaps not as well known in the United States as certain of his illustrious contemporaries, he was nonetheless a remarkably influential painter.

Born in Montreal, Riopelle came to prominence in Paris, where he moved in the same circles as André Breton, Alexander Calder, Joan Miró, and—perhaps most important—the American painter Joan Mitchell, with whom he lived for twenty years. Anyone familiar with Riopelle's work will marvel at the rich colors and sharp textures that characterize much of his work. Riopelle often referred to his paintings as "sculptures in oil," and there can be no doubt that he was a master of his medium, achieving an extraordinary range of effects. He was also extremely prolific, repeatedly producing more than one hundred paintings in the course of a single year.

The Getty Conservation Institute has recently embarked on a major inquiry into the challenges of conserving art of the twentieth and twenty-first centuries. The GCI's Modern and Contemporary Art Research Initiative, in particular, focuses on the scientific study of the materials used by artists over the past one hundred years—concentrating specifically on the analysis of modern paints. One of its primary goals is to ascertain how conservation treatments affect these materials. The initiative also recognizes the importance of disseminating the results of this and related research to others in the field.

Marie-Claude Corbeil, Kate Helwig, and Jennifer Poulin, all working at the Canadian Conservation Institute in Ottawa, have written this book based on their many years of research on the painting materials of Riopelle. The CCI has a long and rich history in conservation research, and it is wonderful to see our institutions working together so effectively. I am delighted that the results of their study can now be shared, and I am confident that this volume will be essential reading for anyone involved in the study or conservation not only of Riopelle's paintings but also of a wide range of modern and contemporary art.

Timothy P. Whalen
Director
The Getty Conservation Institute

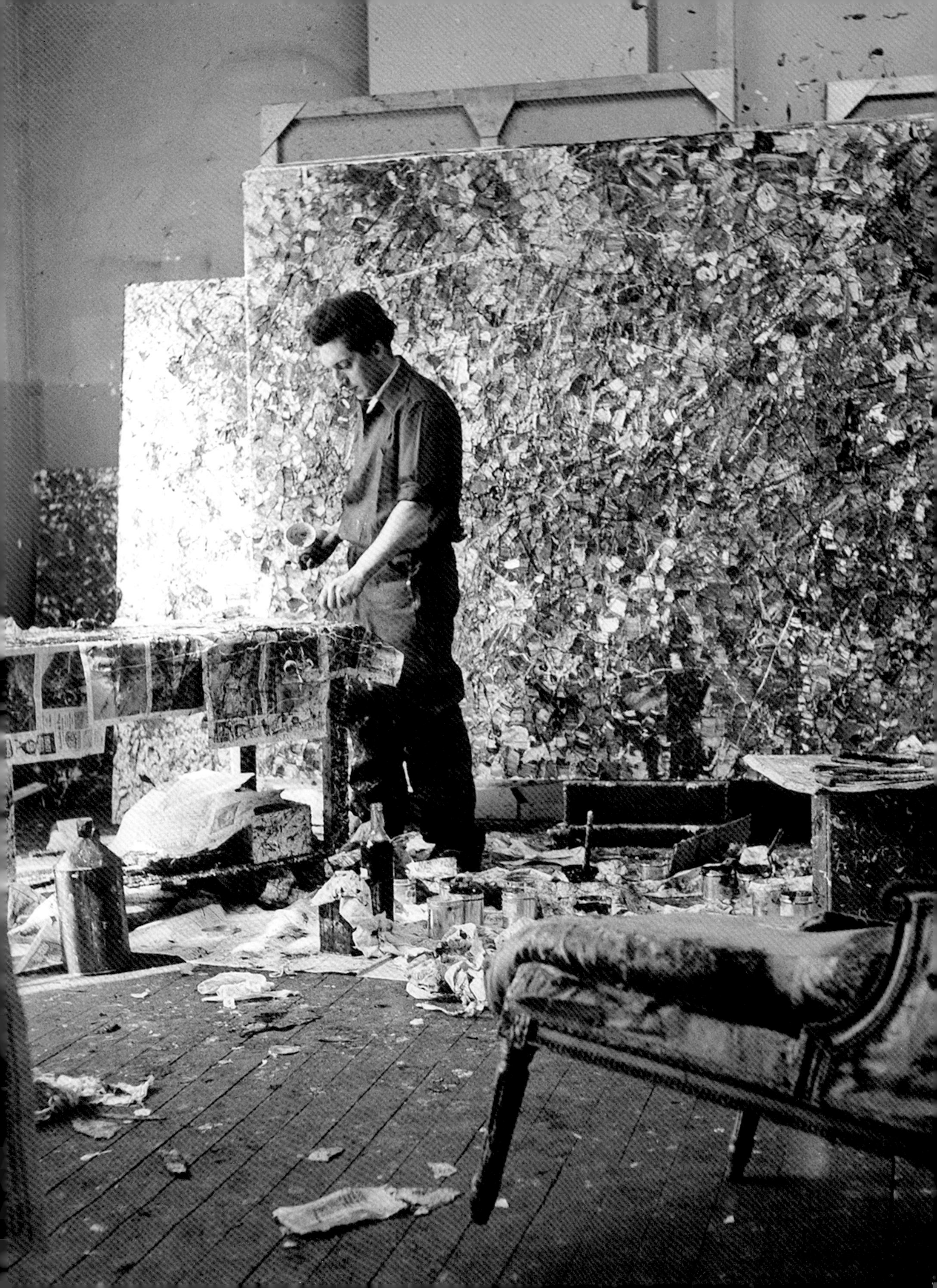

Preface

Riopelle, Alchemist of Matter

Pour que la matière ait tant de pouvoir, il faut qu'elle contienne un esprit.
— GUSTAVE FLAUBERT, *La Tentation de Saint Antoine*

The mind cannot win over matter, for once the mind is asked to do too much, it quickly shows itself to be matter as well.
— PAUL AUSTER, *Moon Palace*

Allons chercher nos images dans l'œuvre de ceux qui ont le plus longtemps rêvé et valorisé la matière : adressons-nous aux alchimistes.
— GASTON BACHELARD, *L'Air et les songes*

My father, Jean Paul Riopelle, used to refer to his paintings as "sculptures in oil," a statement that suggests the importance he accorded to the material, plastic elements of his art. He was, of course, a sculptor as well as a painter, working first in clay, then moving on to wax, sandstone, porcelain, even bread crumbs. In all his work, and in all the many media he explored, he cared deeply about the materials of art, which for him were a constant preoccupation and challenge.

Indeed, throughout his career, Riopelle literally embraced matter. He loved first and foremost the direct and spontaneous contact with the elements, digging in with hands full, without sophisticated tools, except perhaps for the custom-made palette knives that he often used in working with paint. It could be said that for Riopelle the essential and unavoidable task was to acquire knowledge about the materials of his art, the better to master or even transcend them. He was also deeply concerned about the durability and quality of materials, including oil paints, of course, as well as inks, felt-tip pens, and canvases (which he sometimes claimed were of too high quality to withstand the heavy impasto he once used). So, too, did he seek to safeguard and pass on knowledge about artistic materials and techniques. This desire grew as he gained experience over the years, in particular through the printing house Arte, directed by Adrien Maeght, and the engraving and ceramic workshops at the Maeght Foundation in the south of France, where Marguerite and Aimé Maeght, founders of Galerie Maeght in Paris, provided communal work spaces for artists. From this experience emerged Riopelle's desire to create a foundation of his own, a creative space where artists and technicians would collaborate in workshops in sculpture, painting, engraving, and other techniques—a dream that would unfortunately remain unfulfilled.

What can we say about his unlimited imagination? Is research not the "essence" of creation? So much research, so many fascinating experiments! Some remained at the study stage—sets for the Merce Cunningham dance company (1966), plans for work in stained glass (circa 1967), a monumental ceiling for the Montreal town hall (circa 1981), an attempt to combine etching and screen printing (circa 1984), and various architectural designs for the Jean Paul Riopelle Foundation, including a plan for the building that would house the foundation, inspired by an antique sundial at the residence where he lived with his longtime companion, the American artist Joan Mitchell, in Vétheuil, France. Countless other experiments, however, were fully realized, integrating matter in all its forms—solid, liquid, gas. Sometimes this occurred within a single work, as with *La Joute* fountain, in downtown Montreal, which combines water and bronze with the light and energy of fire. The artist's lifelong experimentation produced a massive œuvre of over six thousand paintings, sculptures, and other works, the extent of which is still being established today.

For all these reasons, Riopelle would no doubt have approved of research leading to a rigorous scientific analysis of his œuvre. He would have been thankful for the research being carried out by the Canadian Conservation Institute, research that provides the focus for much of this book. This important work can help prevent the ravages of time and, in some cases, forestall the destruction wrought by man. It will help conserve his œuvre today, as well as provide clues to conservators in the future, and will also be a useful tool, among others, for authentication.

Conservation treatments can be divided into at least three categories: treatments undertaken to save the work, which must be carried out immediately; maintenance procedures, such as dusting and cleaning; and finally various "cosmetic" procedures, such as varnishing. Generally speaking, for Riopelle's paintings, any intervention should be minimal. Varnishing, for example, despite recommendations to the contrary, continues to be an issue with Riopelle's oil paintings. Not only can it damage paint layers; it also shows a lack of consideration for the work and for the artist, as it violates the aesthetic integrity of the original object and hence effectively undermines the artist's intent. (One might question the taste of some collectors who would like to see twentieth-century paintings remain pristine, with no sign of aging such as the fine network of craquelure.) It is important, for example, to respect the artist's exploration of light, which was a constant throughout his oeuvre, notably in the *Jeux de ficelle* (String Games) series, as with the majority of works on paper from 1971 to 1989, which were painted on prints, more specifically proofs and plate-wiping papers. The alternating paint layers create remarkable impressions of light and depth, with the original, underlying print often no longer visible to the naked eye but somehow revealed through the photographer's lens. In other works Riopelle sought different effects, often within a single painting, using both matte paint, for a dull finish, and "Flemish medium," to obtain a more lustrous surface. The textured effect thus achieved is seriously compromised by the application of varnish. Rather than rush in with such expedient and dangerous treatments, how much wiser it is to wait patiently, as in the case of the long-buried army of Xi'an whose terracotta cavalrymen, horses, archers, and foot soldiers attend silently upon the progress of science, which will allow for their careful and respectful conservation, with no irremediable damage done.

In the case of an artist such as Riopelle, conserving the materials of a painting can also preserve our access to the sense of the ineffable that animates it. The critic Pierre Schneider wrote about Riopelle's *String Games*, a series of works on paper (acrylics, mixed media, reworked lithographs) inspired by Inuit practices in which figures having to do with their traditional way of life, such as the seal, the harpoon, and the whale, are fashioned through the manipulation of a string. “In a ceremonial context,” Schneider noted, “the string game becomes propitiation, reconciliation, control. It reminds us of a truth diluted by day-to-day practice: that all language, as a control system, is the rough sketch of a ritual, or a ritual that has lost its sacred aura. We remember words because we give words the power to retain things. Eskimos prefer to play the string game in the fall, to capture the sun in the network of intertwined strings and prevent it from sinking into the sea, putting men in a six-month night” (Schneider 1972). These comments illuminate the fecund spirit of Riopelle’s art as well. As the Inuit people used masks and strings to consecrate the ineluctable rhythms of the seasons, so did Riopelle seek through the manipulation of matter to render homage to a reality that remains forever mysterious, integrating objects familiar and strange—the coal stove shovel, the crucible, the sieve, string, cordage, nails, horseshoes . . .

Riopelle’s love of an immersive contact with the elemental materials of his art explains in part his initial reluctance to use more elaborate techniques such as lithography or etching. Eventually, however, he would master these as well, fashioning a language of mysteriously prophetic clues that leave room for unlimited interpretation. In his final work, *L’Hommage à Rosa Luxemburg*, dedicated to his late companion, Joan Mitchell, Riopelle went so far, in a last attempt to capture the ephemeral, as to create shadows, decoys of migrant birds, figures emblematic of his past life, by spraying them with aerosol paint, as if at the very moment of their last breath . . .

Yseult Riopelle
March 2010

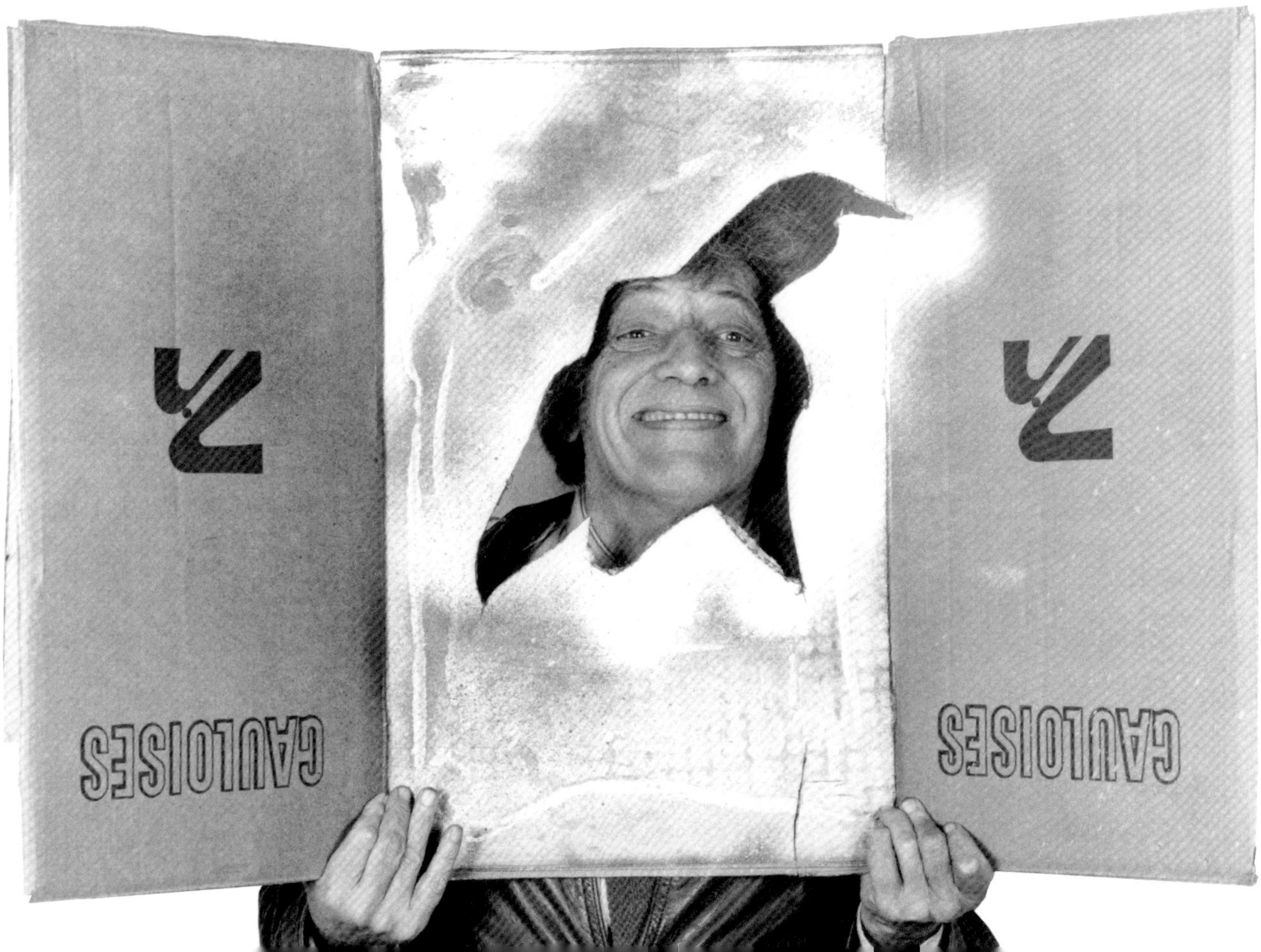

Riopelle holding a stencil he made out of a Gauloises cigarette box, circa 1983.

© Jacqueline Hyde.

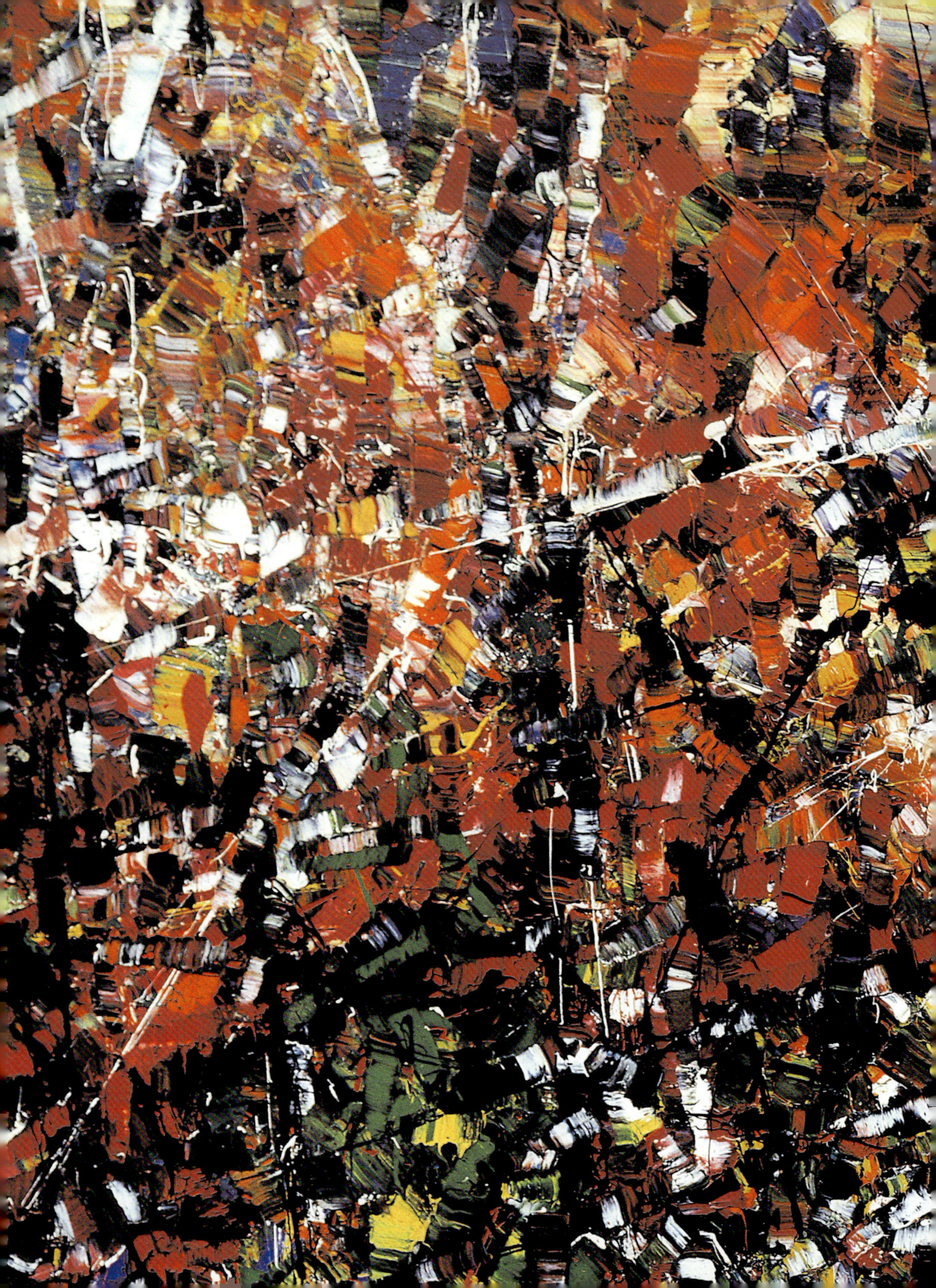

Introduction

In 1968 Jean Paul Riopelle[1] (1923–2002) was introduced by Fernand Seguin, host of the Radio-Canada show *Le Sel de la semaine*, as follows: "Forty-five years old, Riopelle is truly an international painter, one of the ten most important names in contemporary painting, which also means he is one of the ten living painters most sought after. Most of the great museums have acquired his canvases, all international exhibitions worth their salt include his work. In short, he is so famous that people forget he is French Canadian" (Érouart 1995, 60).

Seguin's introduction captured the fascination Riopelle held for people at the height of his career. A giant in the art world, Riopelle not only created an extraordinary œuvre comprising several thousand works but also lived an extraordinary life. He enjoyed racing; he sailed his boat on the Mediterranean Sea; he collected Bugattis. But he enjoyed simple pleasures, too, such as sharing good meals and wine or fishing and hunting with family and friends. Alberto Giacometti, Samuel Beckett, Joan Miró, and Alexander Calder were among his friends; so were the car mechanic and the fly maker. He remained a man of the people even as he achieved fame and success, at a time when very few Canadian artists and even fewer French Canadian artists were known outside Canada. Indifferent to movements and schools, he cared little for fame as long as he had the freedom to develop as an artist and a man, an attitude that somewhat limited the influence of his work later in his career.

Was Riopelle truly a Canadian artist? He left Canada early in his career and painted in France for over forty years, during which there was a period of about fifteen years when he painted in both countries. His last active years were spent solely in Canada. Throughout his life people did not know what to make of him. "In Paris, he was the Canadian or the Québécois, depending on the decade and the attitude of the critics," one scholar has noted. "In New York City, he was identified as a Montreal-born Parisian, a European or a North-American based in Paris. In Canada and in Quebec, his Canadian nationality, or Quebecker identity, was celebrated, with his international reputation always firmly in view. Wherever he was, in fact, Riopelle would always find his artistic identity in an Elsewhere" (Aquin 2002, 34).

Actually Riopelle did not care much about labels. When asked to comment on the fact that New York City was the center of modern painting in the 1960s, he replied, "I don't think there is a centre for painting. There are places where painters meet each other. If we consider French painting, and we say that it's Paris, we still have to admit that not all people doing this painting are French. Far from it. The Cubist school was not particularly French. It's just that they all found themselves in Paris" (Érouart 1995, 62).

Today there is no doubt that Jean Paul Riopelle is one of Canada's most important artists, one of the few whose works are featured in collections worldwide. In light of his stature, a study of his materials and techniques was long overdue, but given his prolific and diverse output, the task seemed virtually insurmountable. Riopelle produced a remarkably wide-ranging œuvre that includes paintings, sculptures, collages, etchings, lithographs, and ceramics, often on a grand scale. To date more than 6,300 works have been catalogued by Yseult Riopelle, the artist's eldest daughter and author of the catalogue raisonné (Riopelle 2010d).

The project undertaken by the Canadian Conservation Institute (CCI), beginning in January 2002, focused on Riopelle's painting—a major field of interest for the artist, comprising over two thousand works (Riopelle 2010d)—with the aim of gathering information to ensure proper care of his paintings and building a database that could be used to detect fakes, several of which have already been encountered.[2] During the course of the project, a large number of paintings were examined, with samples taken from thirty-nine that were representative of Riopelle's work. The earliest painting included in the study was dated 1944; the most recent, 1992. Material details, including the nature of all constituents (such as canvas, ground, and paint), as well as any signatures, inscriptions, or stamps observed on the paintings or on their stretcher frames, were carefully noted and entered into the database. Relevant conservation issues were also investigated. Since Riopelle painted in oil for most of his career, more oil paintings were included in the study than acrylic or mixed-media works, which he started producing in the 1970s. Selected results of CCI's project have appeared in articles aimed at a conservation audience (Corbeil, Helwig, and Poulin 2004, 2006); the overall results are compiled for the first time in this book, where they are augmented by new material. The discussion of materials and techniques, however, does not furnish detailed lists of materials identified in each painting but rather presents a general overview, so as to avoid providing recipes that could be used to create more fakes.

Chapter 1 of this book presents a concise account of Riopelle's life and career, while chapter 2 discusses the art historical context and importance of Riopelle's work. The next two chapters are concerned with the scientific study of the artist's painted œuvre—chapter 3 concentrating on works in oil, chapter 4 on mixed media. The approach employed consisted of careful examination of the paintings and scientific analysis of the collected samples. Others have employed similar procedures to study important artists, from Old Masters to contemporary painters. What proved a challenge in this case was the lack of oral or written sources concerning Riopelle's suppliers, methods, and practices. Riopelle died not long after the project was undertaken, but even if he had been alive there is no guarantee that he would have been forthcoming on these topics, since he never liked talking about the "mechanics" of his painting. No one was allowed in his studio while he was painting, except late in his career, when he became so enfeebled that he required physical assistance. He did occasionally mention technical details in interviews, but, regrettably, his interlocutors did not pursue these comments. We know where Riopelle bought most of his materials when in France; the store closed in the 1990s, however, and the last representative of the family business provided only sketchy, secondhand information, as he had not dealt with Riopelle directly.

Such lack of knowledge along with Riopelle's unorthodox working methods and materials have posed significant issues for conservation. These are explored in chapter 5 in the context of new insights gained from the CCI study. Important considerations related to the treatment of his paintings are outlined as well.

1 A Life in Painting

Born in Montreal on October 7, 1923, Jean Paul Riopelle grew up in a working-class French Canadian neighborhood.[3] His father, Léopold,[4] had started as a carpenter but rapidly prospered in real estate and construction, taking advantage of Montreal's economic growth at the beginning of the twentieth century. By his late thirties Léopold was already living on rental income from his buildings. In 1919, at the age of thirty-seven, he married Anna Riopelle, his second cousin, twenty-nine, who was also well-off, having inherited several properties from her father, a businessman who died prematurely.

Jean Paul was the elder of two boys. His brother, Pierre, died shortly before his third birthday, in 1930, when Riopelle was seven. This tragic event certainly marked Riopelle, who rarely spoke of Pierre, almost as if his brother had never existed, and later described himself as an only child (Érouart 1995, 64). Riopelle, it should be said, did not talk a lot about his childhood in general, admitting that his memories were vague. "I was probably like all the kids in the neighbourhood, maybe a bit more reserved," he remembered, noting that he had been "thrown out of school on occasion" and that, his lack of physical stature notwithstanding, he often got into fights (Érouart 1995, 64).

Financially, Riopelle's childhood was comfortable, but his relationship with his parents, and in particular with his mother, who has been described as demanding and authoritarian, was difficult. His father, a man of his time, was extremely religious. Quebec was then very much under the influence of the Catholic Church, which controlled such institutions as labor unions, professional associations, and agricultural cooperatives. It also held sway over education, despite attempts by politicians and businessmen to reform the traditionalist school system, and imposed its point of view on all aspects of cultural life, including fashion, literature, and art. While members of artistic circles in Montreal had stayed abreast of developments in Europe through visiting artists (Nasgaard 2009, 7), the population of Quebec in general was largely isolated from a Europe whose cultural landscape in the decades following World War I was being transformed by the upheavals of modernism.

Perhaps most important, the church had strong ties to politics, a situation that culminated in the governments of Maurice Duplessis, a fervent Catholic, who was premier from 1936 to 1939 and from 1944 to 1959. While Duplessis made important contributions to the economic and social development of Quebec, he showed disdain for civil liberties, as exemplified by his strong-arm tactics with striking unions and the so-called Padlock Act, which allowed the closing for one year of any building and the confiscation of any printed matter deemed to have been used for "propagating communism or bolshevism." Anyone publishing or distributing such material could be imprisoned without appeal for up to a year. The Duplessis years were referred to as the Grande Noirceur (Great Darkness) by his opponents, a term that captures the conservative and oppressive ambiance that prevailed when Riopelle was a teenager and young adult.

Artistic Beginnings

Riopelle had his first encounter with art at a young age,[5] when he started taking drawing and painting lessons with Henri Bisson. Bisson, a graduate of the École des beaux-arts in Montreal, was an academic painter steeped in the realist tradition. Riopelle took private lessons with him every weekend and painted in Bisson's studio or outdoors, in the countryside. "He was a marvelous teacher for the very good reason that he did his painting... in his corner, and let me do mine," Riopelle later recalled. "The corrections came afterwards." In those years, Riopelle remembered, "painting was my passion" (Érouart 1995, 65–66).

Also inclined toward mathematics, Riopelle decided to enroll at the École polytechnique de Montréal in September 1942, after a year of preparatory school. He apparently did not focus on painting as intensely as before, however; after he got into Polytechnique, he "partly let it drop," he later said (Érouart 1995, 66). Not surprisingly, perhaps, he left the school after one year. He took correspondence courses in architecture and evening classes at the École des beaux-arts but was disappointed because "all students did was to copy statues" (Érouart 1995, 66). "I was fairly quick and knew it," he remembered. "I only had to attend one session a month, and then take a test the next day in order to pass. It wasn't very real to me. I didn't want to wait all those years to finally be a painter. I'd been painting for a long time. I should say that I was an academic painter" (Érouart 1995, 66). His work during this time consisted mainly of landscapes and still lifes (fig. 1). Riopelle was not the only one to dislike the conservative leadership and curriculum of the École des beaux-arts; many students criticized the school, and, like him, a number quit (Ellenwood 1992, 6).

In 1943 Riopelle enrolled at the École du meuble. While not, strictly speaking, an art school, as it taught applied arts such as industrial drawing and design, the École du meuble was in those years a fertile ground for artistic innovation. Among his professors was Paul-Émile Borduas, already an important figure in Canadian art. Although Borduas's artistic beginnings were quite traditional, he became an important advocate for the arts, and especially modern art, when he joined the École du meuble in 1937 and was elected vice president of the Contemporary Arts Society in 1939 (Nasgaard 2009, 8–9). Around the same time, the outbreak of World War II had brought the return to Canada of the painter Alfred Pellan, who had been living in Paris since 1926 and was deeply engaged with European movements such as Cubism and Surrealism. A number of Europeans who would further the cause of modern art—both artists and such nonartists as the Dominican priest Father Marie-Alain Couturier—sought refuge in Canada or the United States (Nasgaard 2009, 9). With them came an influx of new ideas that probably contributed to Borduas's evolution from figurative painting to abstract painting to what would become the Automatist movement.

Automatist painting was rooted in Surrealism, particularly in André Breton's concept of automatic writing, a "stream-of-consciousness technique that, when faithfully observed, would not only tap into the subconscious but also liberate artists from obligations to content or representation, and absolve them of rational or moral inhibition" (Nasgaard 2009, 13). These concepts had flourished in Europe after World War I but were still relatively new in Quebec in the 1930s. Automatist painting, as defined by

FIGURE 1
Jean Paul Riopelle, *Nature morte au violon*, 1941, oil on canvas mounted on cardboard, 35.8 × 27.5 cm. Montreal Museum of Fine Arts, inv. 2004.142. Gift of Martin Gauvreau. Catalogue raisonné no. 1941.003H.1941.

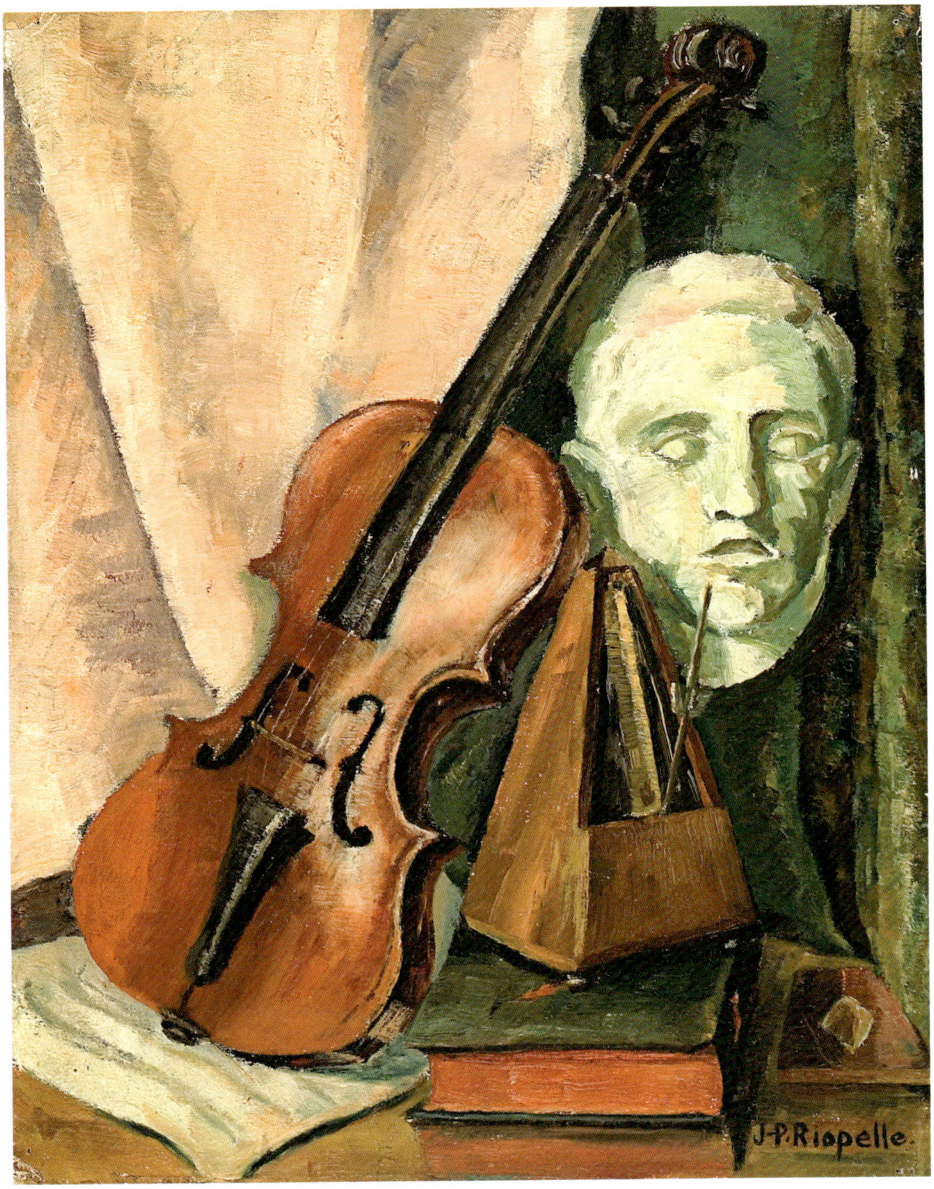

Borduas, was "unpremeditated writing in plastic matter. One shape calls up another until a feeling of unity is achieved, or a feeling that to go further without destruction is impossible. During the process, no attention is paid to content. This freedom is justified by the conviction that content is inevitably linked to form" (Borduas 1985, 46).

First shocked by the way Borduas looked at art, Riopelle, who did not think much of "so-called modern painting," initially provoked his teacher. In his 1968 interview with Fernand Seguin, he recalled how Borduas reacted: "That was his great virtue: instead of writing me off, he saw that my antipathy could be turned into something else. And as he was a perfect guide, I changed my attitude in less than a year" (Érouart 1995, 67). When asked if Borduas influenced his painting, Riopelle answered, "Where painting is concerned, I don't think Borduas wanted to influence people. He wanted to free them, to help them find themselves" (Érouart 1995, 70).

At the École du meuble Riopelle joined a group of young artists who, led by Borduas, became known as the Automatists. The group's members included painters, dancers, poets, and playwrights who were enthusiastically exploring new artistic grounds. The importance of the Automatists in the history of modern art is now well established. One art historian has gone so far as to state that "the first lyrical abstract group in the world was neither Parisian, or from New York City, but was Québécois."[6] Riopelle shared a studio with two painters from the group, Marcel Barbeau and Jean-Paul Mousseau. His first abstract oil paintings and ink-and-watercolor works on paper date to 1945.

At that time, after World War II, the art scene in Europe and North America was undergoing important and dynamic changes. Abstract movements opposed to Cubism and Surrealism as well as to geometrical abstraction, such as Expressionism and Lyrical Abstraction, were emerging in New York and Paris, and Riopelle was eager to measure himself against new concepts and ideas. In 1946, along with several other painters from the group, he participated in an exhibition at a dance studio in New York City; apparently, however, it did not generate much interest (Nasgaard 2009, 44). The same year he made his first trip to France. In October he married Françoise Lespérance, the sister of a close friend. The young couple would return to France several times over the course of the next two years, and it would be in France that the artist would come into his own.

Rise to Prominence

Paris in the postwar years was a center of intellectual and cultural ferment. The young Riopelle, in his early twenties, immersed himself in the vibrant artistic scene, quickly becoming acquainted with a number of artists, art dealers, critics, and other influential people who later played important roles in his career. In 1947 he met the dealer Pierre Loeb, who introduced him to several Surrealist painters and writers, in particular André Breton, who had been the inspiration for the Automatist movement in Quebec. During the summer Riopelle participated in several exhibitions in Paris, including *Automatisme*, at the Galerie du Luxembourg, with fellow Automatist painters Borduas, Barbeau, Roger Fauteux, Fernand Leduc, and Mousseau. He frequented the Surrealists, who were divided along ideological lines, with one group loyal to Breton and the other linked to the French Communist Party and highly critical of Breton's equivocal attitude toward communism. This schism would lead to Breton's break with the Communists in the form of the manifesto *Rupture inaugurale,* which Riopelle cosigned. Breton invited Borduas and Riopelle to participate in the exhibition *Le Surréalisme en 1947*, presented at Galerie Maeght, but Borduas declined, leaving Riopelle the only Canadian artist represented.

At the end of this fruitful summer Riopelle was back in Canada, where in early 1948 his first daughter, Yseult, was born. He spent the year in Montreal and Saint-Hilaire, playing an important role in the creation of one of the seminal documents of modern Quebec. Launched at the Librairie Tranquille in Montreal on 9 August 1948, *Refus global* was an antiestablishment and antireligious manifesto rooted in the traditions of the European avant-garde. The main essay was written by Borduas and signed by fifteen members of the Automatist group, including Riopelle (Nasgaard 2009, 53).[7] Bruno Cormier (then a poet and playwright), the poet Claude Gauvreau, the painter Fernand Leduc, and Françoise Sullivan (then a dancer) wrote the other texts in the manifesto, and various group members contributed illustrations. Riopelle apparently played a key role in its genesis by introducing his Canadian colleagues to the French Surrealist manifesto *Rupture inaugurale*. He had signed that document but also strongly favored the creation of a manifesto that would address issues specific to Quebec (Bernier 1997, 30).

Refus global defiantly rejected the traditional values of Quebec and advocated a new collective hope, constructed from acts of passion, that would shape the future. It was premised on "total risk through global refusal" (Borduas 1985, 38). "We must break with the conventions of society once and for all, and reject its utilitarian spirit," the manifesto provocatively declared. "We must refuse to function knowingly at less than our physical and mental potential.... We refuse to be confined to the barracks of plastic arts—it's a fortress, but easy enough to avoid. We refuse to keep silent.... We will not accept your fame or attendant honors. They are the stigmata of shame, silliness and servility.... We reject all forms of INTENTION, the two-edged, perilous sword of REASON" (Borduas 1985, 37). In terms of art specifically, the rejection of intention and reason was very much in line with the Surrealist project of freeing expression from the constraints of an a priori conscious intent to create an art that was free to discover itself in the course of its own making. In terms of society generally, the manifesto was a clarion call for change: "We foresee a future in which man is freed from useless chains, to realize a plenitude of individual gifts, in necessary unpredictability, spontaneous and resplendent anarchy. Until then, without surrender or rest, in community of feeling with those who thirst for better life, without fear of setbacks, in encouragement or persecution, we shall pursue in joy our overwhelming need for liberation" (Borduas 1985, 41).

This manifesto is said to have paved the way to what is known in Quebec as the Quiet Revolution, a period of rapid change in the 1960s characterized by open debate that led to important reforms. Its publication, however, had some unfortunate immediate consequences, not the least of which was Borduas's expulsion from the École du meuble, as the school declared him unsuitable for the kind of teaching it wished for its students. A war of words in newspapers and magazines followed the manifesto's publication. Catholic intellectuals strongly condemned it, while Automatists and other opponents of the Duplessis government defended it and protested the firing of Borduas.

Later that year Riopelle left Montreal with his family—which would soon include a second daughter, Sylvie—to settle in Paris, first in the suburb of Bois-Colombes and then on avenue Saint-Mandé in the capital's 12th arrondissement. Riopelle's departure from Quebec was motivated not by the controversy occasioned by *Refus global* but by France's appeal as a place for him to live and work. Indeed, over the next several years, he continued to work assiduously, concentrating increasingly on oil painting. He had his first solo exhibition in Paris in 1949, at La Dragonne (Galerie Nina Dausset); the exhibition catalogue included a preface by André Breton, Élisa Breton, and the Surrealist poet Benjamin Péret. It was in this preface that Breton referred to Riopelle as a "master trapper" (*trappeur supérieur*), a name that would follow him all through his career. This first solo exhibition, although an important milestone in Riopelle's career, did not bring the expected success. Riopelle's works were exhibited more and more widely, appearing in various Parisian galleries in 1950 and 1951 and at the São Paulo Biennale in 1951; sales, however, continued to be rare.

Then, in 1952, his perseverance finally paid off when the sculptor Isabelle Waldberg asked Riopelle to exhibit his paintings with her sculptures at Galerie Henriette Niepce. "The day of the *vernissage*," Riopelle later recalled, "a dealer walked in and said to me: 'I very much like your work.' I looked at him and I said: 'Listen, you've been telling me that for seven or eight years.' 'No, but I am serious, I'll buy everything,' [he said]. I said: 'That's bizarre! But if that's the way it is, so much the better.' Then one of his friends came up to me and said: 'Monsieur, do you need a studio?'" (Érouart 1995, 76). The dealer was Pierre Loeb; in addition to purchasing Riopelle's entire lot, he offered

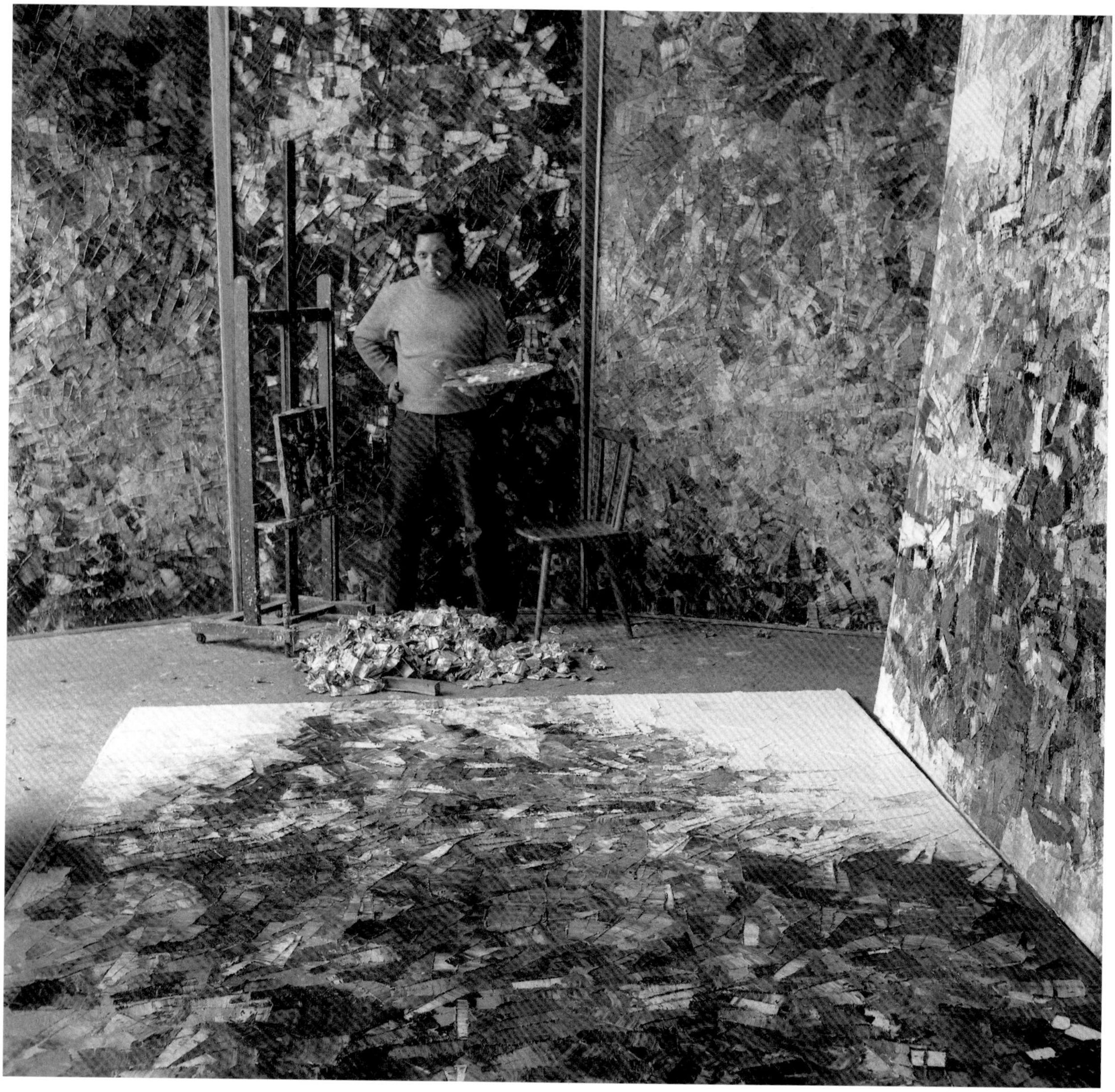

FIGURE 2
Riopelle in his studio in the Paris suburb of Vanves, 1956.

© Atelier Robert Doisneau.
Photo: Robert Doisneau/Gamma Rapho

to sign the artist as a client.[8] The studio loaned by Henry Fara, one of Loeb's friends, meanwhile, would serve Riopelle well. "[Fara] said, 'I'm leaving for Egypt tomorrow, come to 52 rue Durantin, the key is there, I'll be back in six months,'" Riopelle remembered, adding, "I stayed three years in that studio" (Érouart 1995, 76). Riopelle's career was launched. From this point on, his work would meet with immediate and continuous acclaim (fig. 2).

The Artist and His Circle

Riopelle's success in establishing himself as an artist can be attributed to his talent and efforts, of course, but his personality and the spirit that prevailed in Paris in the late 1940s and early 1950s played a large part as well. As one foreigner among many in postwar Paris, Riopelle joined a company of young expatriates, many of whom would become prominent and influential artists. Other members of the Automatist group, like Fernand Leduc and Marcelle Ferron, established themselves in the French capital, with the group losing its initial cohesion as its members pursued independent careers. Many Americans, too, were in Paris at this time. Among them was Sam Francis, who introduced Riopelle to the American group that included Ellsworth Kelly, Shirley Jaffe, and Norman Bluhm.

In Riopelle's circle of friends were writers, other painters, and critics. One was the writer and art critic Georges Duthuit, Henri Matisse's son-in-law, who wrote a seminal article—the first of many—on Riopelle in 1952 (Duthuit 1952). It was translated into English by Samuel Beckett (Bernier 1997, 37), who at the time was doing translation to earn money and whose landmark play *En attendant Godot* would premiere in Paris the following year.[9] Duthuit introduced Beckett to Riopelle, and they became friends. It was also through Duthuit that Riopelle met the art dealer Pierre Matisse (Henri Matisse's son and Duthuit's brother-in-law), who became Riopelle's New York dealer in 1953. The art critics Pierre Schneider and Patrick Waldberg were among his friends as well. He also grew close to theater people, musicians, and sculptors, as well as people outside the artistic milieu. "I've had friends, great friends, and whether they're famous or not is of no importance," Riopelle recalled. "It could be [well-known artistic and literary figures], or it could be a car mechanic, someone I met at a café" (Érouart 1995, 7).

His friends also included dealers and collectors. Riopelle would have several dealers in Paris: Pierre Loeb would be followed in 1956 by Jacques Dubourg, then in 1966 by Aimé Maeght. And he had dealers abroad—Pierre Matisse in New York, Arthur Tooth in London. "I was always close to my dealers because our relationship was always based on shared 'opinions,'" Riopelle once said. "When my dealers' opinions changed, I changed dealers." But in the same breath he added: "Emotional bonds have always been important to me. Once they're violated, I make a clean break. And I don't forgive. It's the same when my friends are attacked: I don't pardon the aggressor" (Érouart 1995, 29). He went so far as to say that the price of paintings was related to "the talent of the dealers. Nothing else" (Érouart 1995, 6). As for collectors, Riopelle noted: "At the beginning,...I was not thought of in terms of my 'value'...but simply as...someone it would be good to know. As a human being rather than a financial investment." Collectors, he said, "were just friends" (Érouart 1995, 29).

One of Riopelle's most important relationships was with the American artist Joan Mitchell, whom he met in 1955. Riopelle started living with her in 1959 after separating from his wife, Françoise. The relationship between Riopelle and Mitchell, which would last until 1979, was passionate and tumultuous. It was also complex, probably because they were both important artists, which led to emulation but also to rivalry. In the United States, where they spent summers in East Hampton in the early 1960s, Mitchell was very much in her element, but France was Riopelle's territory.

Artists, dealers, collectors, family, and friends would continue to play important roles throughout Riopelle's life, along with the other things that made life enjoyable for him—travel, food, wine, sailing on the Mediterranean. All these were intertwined with his painting. Riopelle gave a few of his paintings to Pierre Matisse, for example, in exchange for a sailboat, noting that "my sailboat is my painting" (Érouart 1995, 6). In addition to their passion for sailing, Riopelle and Matisse shared an interest in cars, and Riopelle developed an impressive collection of vintage models. In 1978 Baron Philippe de Rothschild asked him to design the label of his prestigious wine Château Mouton Rothschild,[10] for which he was reportedly paid twelve crates (144 bottles) of the baron's great Bordeaux (de Billy 1996, 229). In 1974 Riopelle bought a general store

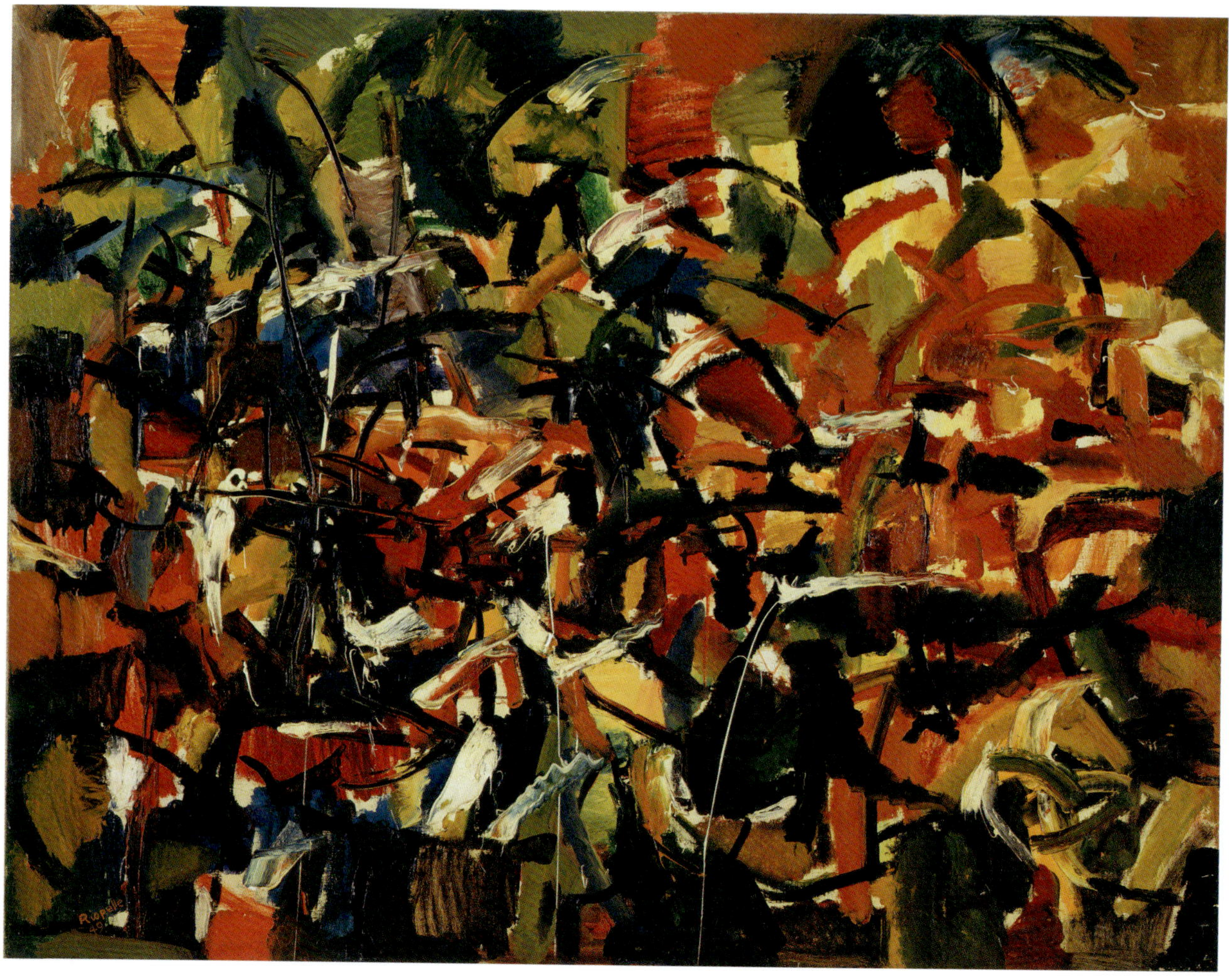

FIGURE 3

Jean Paul Riopelle, *Le Perroquet vert*, 1949, oil on canvas, 110.2 × 140 cm. Musée national des beaux-arts du Québec, inv. 69 36. Catalogue raisonné no. 1949.010H.1949. Purchase.

Photographer: Jean-Guy Kérouac

in Sainte-Marguerite-du-Lac-Masson, north of Montreal, with his friend Champlain Charest. They turned the store—which dated from 1859—into a restaurant, Le Va-nu-pieds. It was a natural venture for Riopelle, given his interest in food and wine. Renowned today for its fine food and one of the best wine cellars in North America (thirty-five thousand bottles), the restaurant, now called Le Bistro à Champlain, also houses an important art collection that includes, of course, many works by the former owner.

The Exploration of Matter

Over the course of nearly fifty extremely prolific years, as he fashioned an oeuvre of remarkable depth and scope, Riopelle explored the materials and techniques of art. This profound and abiding interest may well have had its roots in the early influence of Automatism. Borduas had invited his students to think "painterly" rather than "literally," to concentrate on the act and the material components of painting, elevating matter to be "the essential carrier of a painting's meaning" (Nasgaard 2009, 13). Throughout his career Riopelle would continue to discover materials and techniques that allowed him to express himself in new ways.

While Riopelle's early oil paintings had much in common with his ink-and-watercolor works on paper, the black paint motifs and lines of the former (fig. 3) echoing the delicate black ink patterns of the latter, the similarity would disappear as Riopelle became increasingly interested in the possibility of oil paint as matter. The paintings gained texture, and lines developed from a few splatters to a complex network superimposed onto a polychromatic background. Paint was applied thickly and generously in a riot of color.

In 1953 Riopelle began the *Mosaïques* series, probably his most famous works. In these monumental oils, which seem almost closer to murals than to easel paintings, the sharply juxtaposed strokes of the palette knife created shattered, virtually kaleidoscopic color motifs (fig. 4). The initial explosion of color was gradually invaded by white, creating luminous zones (fig. 5); in later paintings the shattered color motifs were replaced by more elongated strokes.

During the 1950s Riopelle produced an amazing number of works, often more than one hundred oil paintings in a single year, not to mention an impressive number of works on paper (ink paintings, watercolors, and gouaches). He continued to work at a phenomenal pace in the next two decades, producing between thirty and more than one hundred paintings each year. He worked largely in oil but began to explore other media as well. In the early 1960s he added to his range of media by producing pastels, charcoals, and sanguines and starting to use felt-tip pens. While stuck indoors at a ski station in the Pyrenees after a knee injury during the 1964–65 winter holiday, he resorted to materials that were made available to him there, mixing watercolors, inks, and pastels on either paper or cardboard supports and developing techniques he would more fully explore in his later mixed-media works (Brunet-Weinmann 2009, 35). Stylistically, too, his work continued to evolve, with more explicit references to nature and more easily recognizable shapes. The monumental painting commissioned by the Government of Canada for the Toronto airport, *Point de rencontre* (fig. 6),[11] for example, is a clear step toward figurative painting.

In 1966 Riopelle began his association with the art dealer Aimé Maeght, who owned Galerie Maeght. Aimé Maeght and his wife, Marguerite, had established the Maeght Foundation in 1964,[12] a communal work space (as well as a lovely museum)

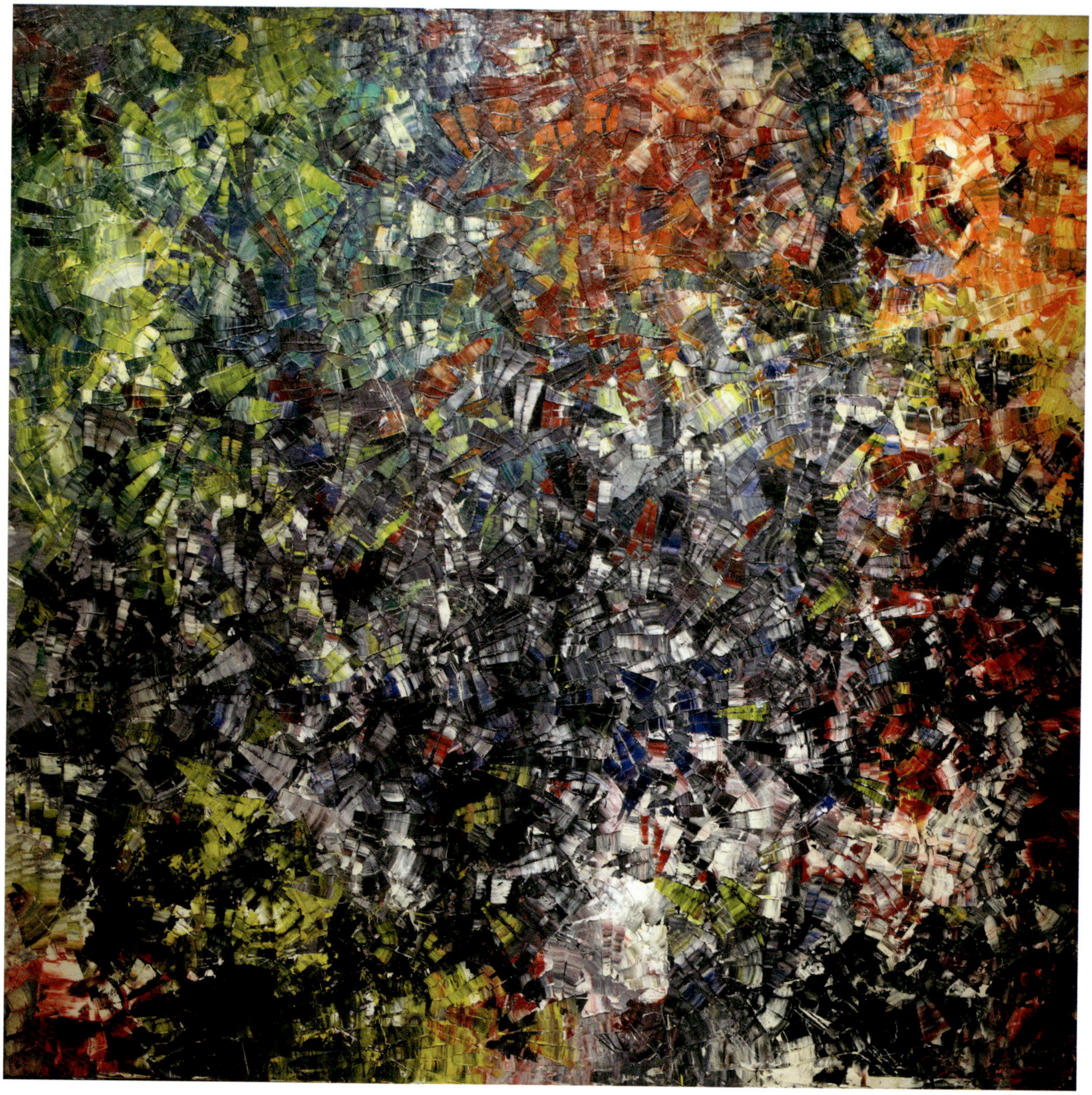

FIGURE 4
Jean Paul Riopelle, *Chevreuse II*, 1953, oil on canvas, 300 × 300 cm. Art Gallery of Ontario, inv. 2003/1664. Catalogue raisonné no. 1953.030H.1953.

FIGURE 5
Jean Paul Riopelle, *Rencontre*, 1956, oil on canvas, 100 × 81 cm. Ludwig Museum, Cologne, inv. ML 76/2998. Catalogue raisonné no. 1956.001H.1956.

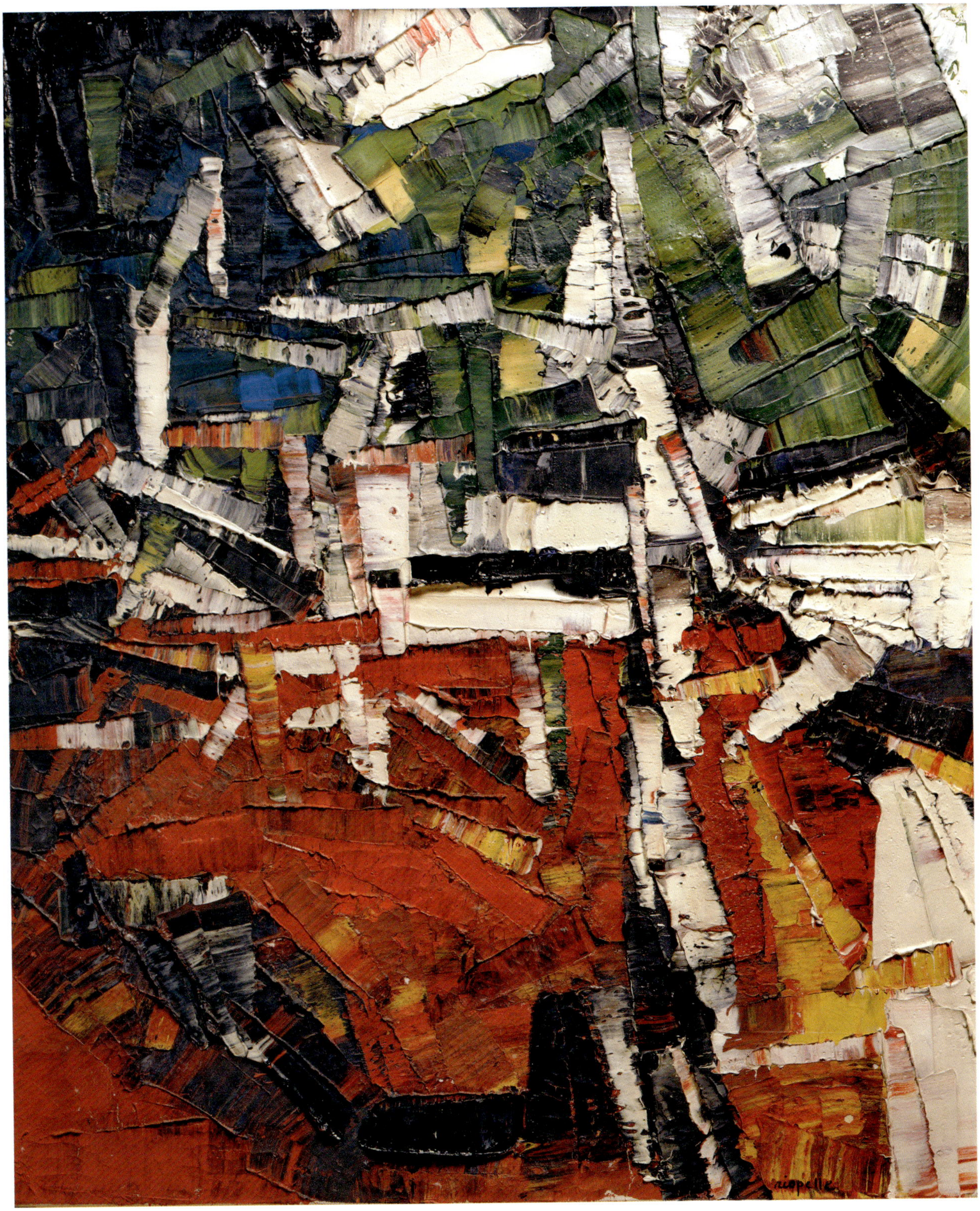

FIGURE 6
Jean Paul Riopelle, *Point de rencontre*, 1963, oil on canvas, 428 × 548 cm (3 vertical canvases of 286 × 182.7 cm on top of 2 horizontal canvases of 142 × 274 cm). Fonds national d'art moderne (France) (attributed to Paris Bastille Opera House), inv. 90069. Catalogue raisonné no. 1963.005H.1963.

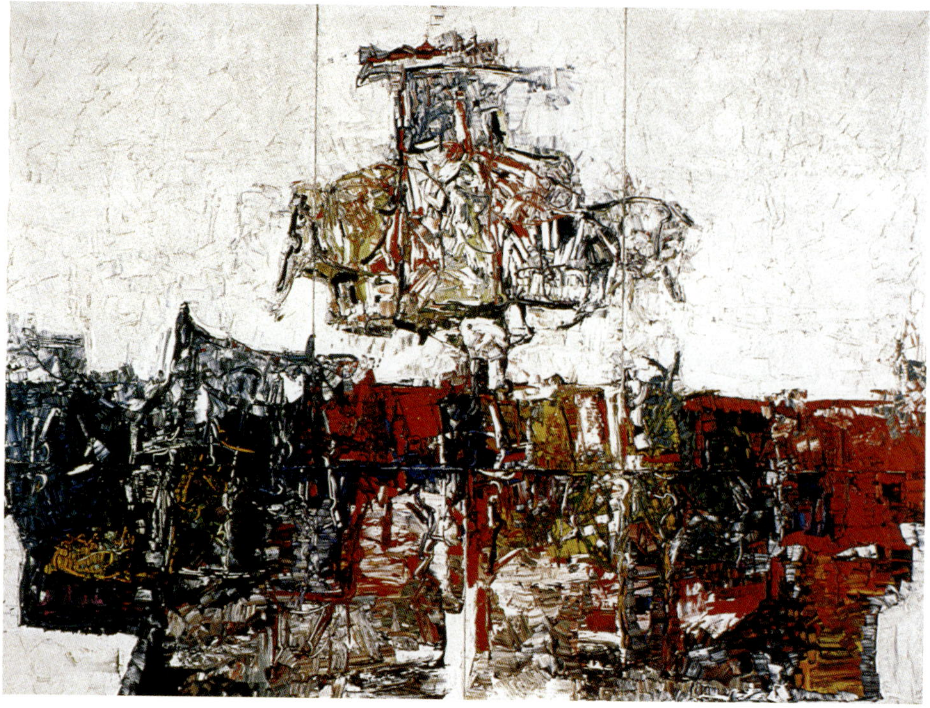

near Saint-Paul-de-Vence where artists had access to workshops and studios and were encouraged to experiment. The lithographs and etchings that Riopelle began producing at Adrien Maeght's printing firm, Arte, where he worked alongside Alexander Calder and Joan Miró, marked an important turn in his artistic career. Exploring new media, however, did not mean that he neglected his painting, as in 1966 he produced about seventy-five oil paintings. In 1967 he began to create lithograph collages, using proofs and plate-wiping papers from the first lithographs he produced at Arte. One reason he took up the collages had to do with the difficulty and expense of obtaining other materials. "Collage," Riopelle would later say, "costs less than tubes of paint" (Érouart 1995, 54).

The early 1960s also saw a renewal of his interest in sculpture, an art he had practiced on a small scale early in his career,[13] and he acquired a foundry in association with the sculptor Roseline Granet. In 1969 he started work on elements of his most ambitious sculptural project, *La Joute* fountain. A plaster version was exhibited at the Maeght Foundation at the end of 1970. The bronze was cast in Italy in 1974 and was installed at the site of the 1976 Montreal Olympic Games. Now relocated in downtown Montreal at Place Jean-Paul-Riopelle, in front of the city's convention center, *La Joute* is one of Riopelle's very few public art pieces.

Meanwhile, Riopelle continued experimenting with different materials and techniques. He used acrylics for the first time in 1971, for the series *Jeux de ficelle* (String Games), in which clearly figurative elements such as strings and knots were used to produce compositions that nevertheless remained abstract. In 1977 he started to work with silverpoint. Two years later he turned to a whole new project: a large wall composed of sixty-three elements combining porcelain and grogged stoneware linked by ropes. Titled *Le Mur (hiboux liés)*, the wall was done with the ceramist Hans Spinner over a two-year period in the studios of the Maeght Foundation. The combination of translucent porcelain and matte stoneware allowed Riopelle to create interesting plays of light, as he did in his paintings, sculptures, and works of art on paper. He would work again with Spinner in 1984 and 1985 on compositions of glazed lava stones.

FIGURE 7
Jean Paul Riopelle in his studio in Saint-Cyr-en-Arthies, in front of *Le Bestiaire*, 1989.

Photo © Jacqueline Hyde.

FIGURE 8
Jean Paul Riopelle, *L'Oie hélico*, verso of *Matinée au Cap Tourmente* (*Les Faisans dans la volière*), 1990, mixed media on wood panel, 124.5 × 246.5 cm. Montreal Museum of Fine Arts, inv. 2001.165.2. Purchase with a special grant from the Government of Quebec. Catalogue raisonné no. 1990.228.1990.

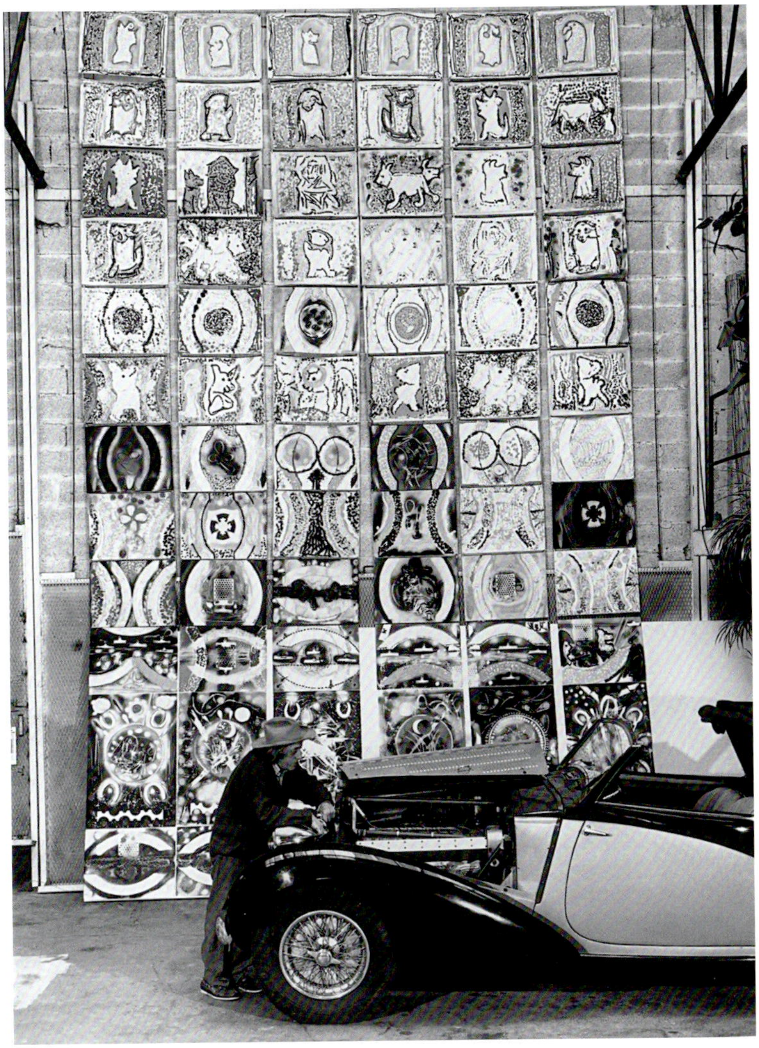

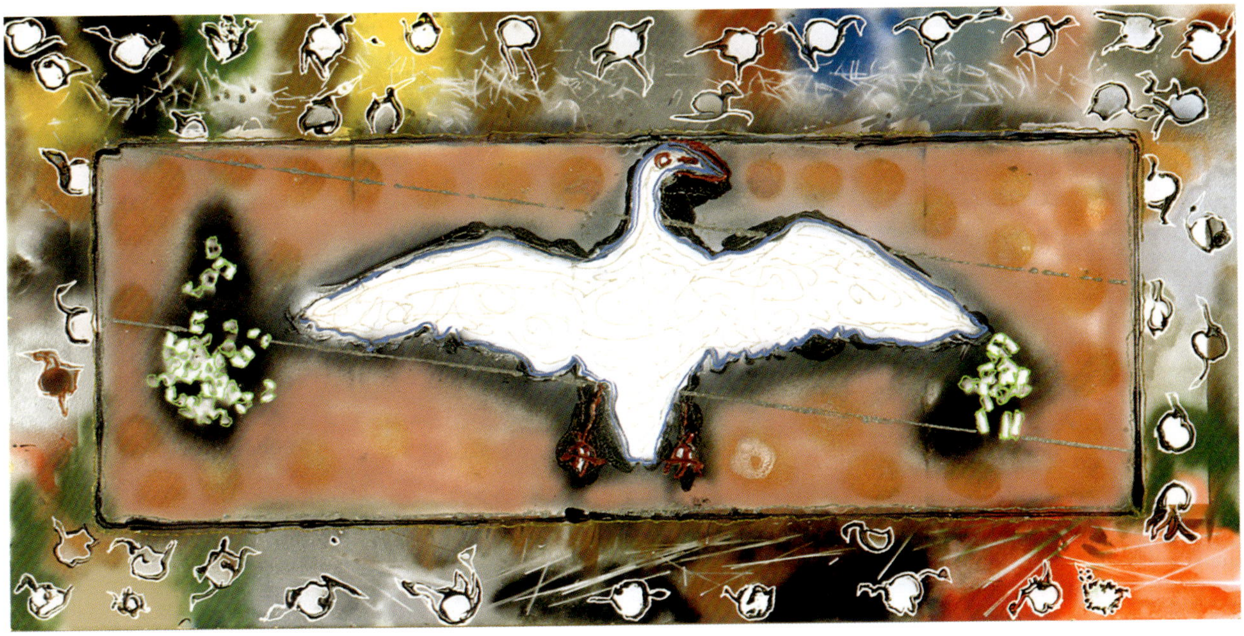

FIGURE 9
Jean Paul Riopelle, *Pangnirtung*, 1977, oil on canvas, triptych, 200 × 560 cm (3 canvases; from left to right: 195 × 130; 200 × 300; 195 × 130). Musée national des beaux-arts du Québec, inv. 97.113. Purchase with a gift from Loto-Québec. Catalogue raisonné no. 1977.211H.1977.

Photographer: Patrick Altman

In the late 1970s and 1980s Riopelle's health began increasingly to affect his artistic choices. He had shown the first symptoms of osteoporosis in 1968, a condition that would gradually worsen. In 1977 a serious knee injury limited him for a time to small-scale works. In 1979 he stopped painting in oil, largely because he no longer possessed the necessary physical strength, and worked primarily on paper for most of the next decade. (Often these works on paper were later adhered to canvas for conservation reasons or to reproduce the structure of a canvas painting.)[14] Riopelle broke several vertebrae in a 1986 car accident and two years later spent a year in the hospital and a rehabilitation center because of new fractures to his spinal column. The following year, however, despite his fragile condition, he executed over 350 works on paper, including one of his largest works, *Le Bestiaire*, composed of seventy-two works, 6.55 meters high by 4 meters wide (fig. 7). He continued to produce lithographs, felt-pen drawings, and collages and began creating silkscreens. The late 1980s brought an increased use of mixed media, including aerosol paint as well as paint for crafts and fabrics (fig. 8). The geese motif, which first appeared in 1976 in a series of drawings, became dominant in 1981; it would remain a principal thematic preoccupation for the rest of his career.

Painting Places

While many of his works are simply titled *Composition* or bear no title at all, many titles evoke places Riopelle visited in his travels (*Autriche*, *Espagne*, *Arles*, and so on). Always curious about people and places, he had started traveling in his youth. As a

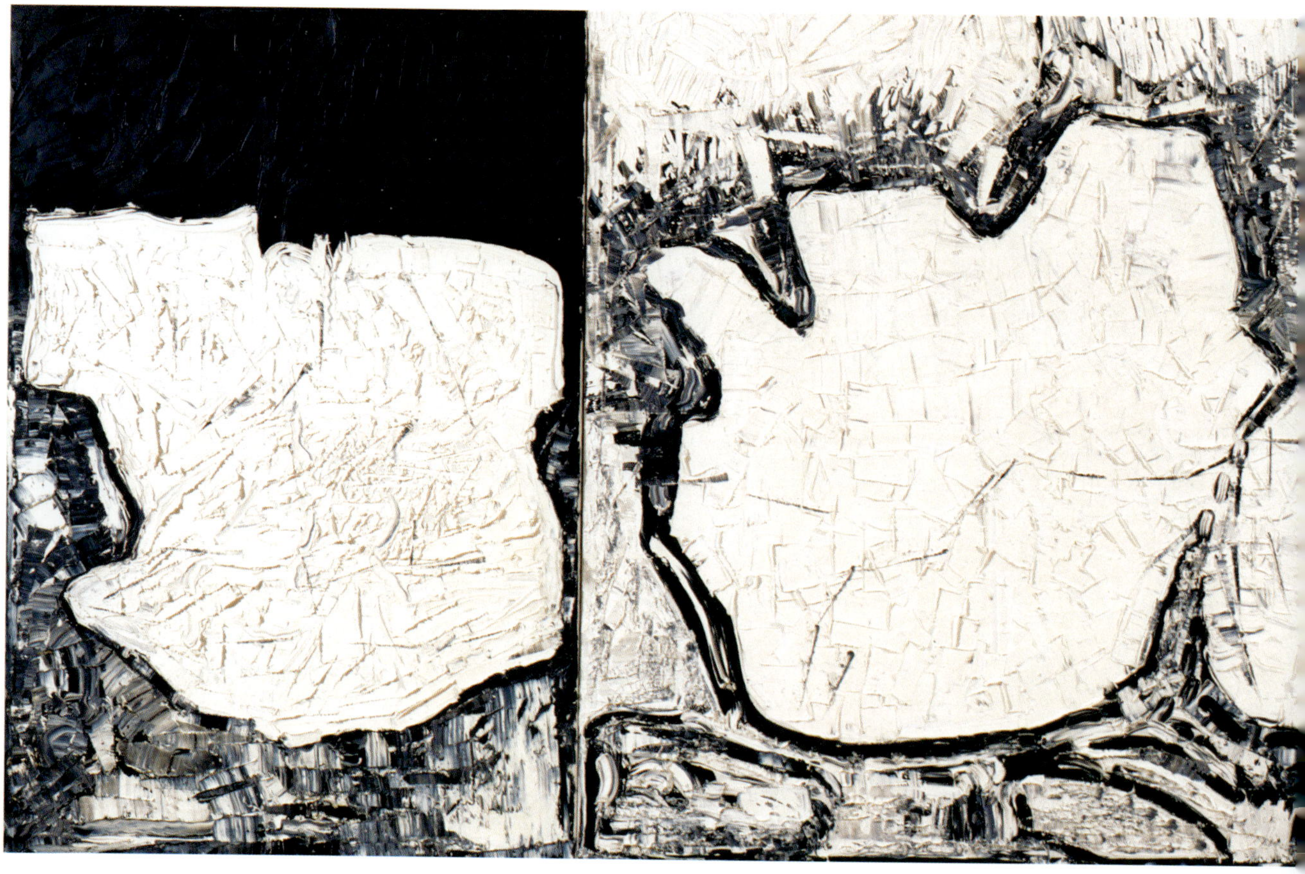

teenager he enjoyed spending his summers in Saint-Fabien-sur-Mer with his friend Jean Lespérance (and Jean's sister Françoise, whom he would marry), and as a young artist, of course, he had traveled to New York and Paris. Once settled in France, he traveled widely in Europe with family and friends. For a time he returned to Canada infrequently, but he began to make longer trips in the late 1960s and early 1970s, spending a great deal of time fishing and hunting, mostly in Quebec. Riopelle also explored the wilderness of Cap Tourmente, the Laurentians, Abitibi, and James Bay and traveled to the far north and the Arctic, journeys often reflected in his paintings. The series *Rois de Thulé* and *Icebergs,* for example, were inspired by his trips to the far north. The latter, comprising about two hundred works, painted in 1977, were his last large-scale oil paintings in a series (fig. 9).

Riopelle's studios, where he almost always painted alone, were also places of essential importance. In France he had moved his studio a number of times. After working in the studio on rue Durantin in Montmartre that was loaned to him, he moved in 1954 to a studio in Vanves, in the suburbs of Paris. In 1969 Riopelle rented a hangar in Saint-Cyr-en-Arthies, near Vétheuil, first to house his car collection and later to include his studio. The hangar was close to where he lived with Joan Mitchell starting in 1967, on a property where Claude Monet once had his studio. Riopelle purchased the hangar in 1974 and began living there in 1979, after his separation from Mitchell. In Canada he designed a studio-home in Sainte-Marguerite-du-Lac-Masson, north of Montreal, in 1974, based on a concept that he had submitted to the École du meuble as a student. Built on a slope, on the bank of a lake, the two-story studio-home housed living spaces on the first floor and the studio on the second floor, at road level, so moving artworks in and

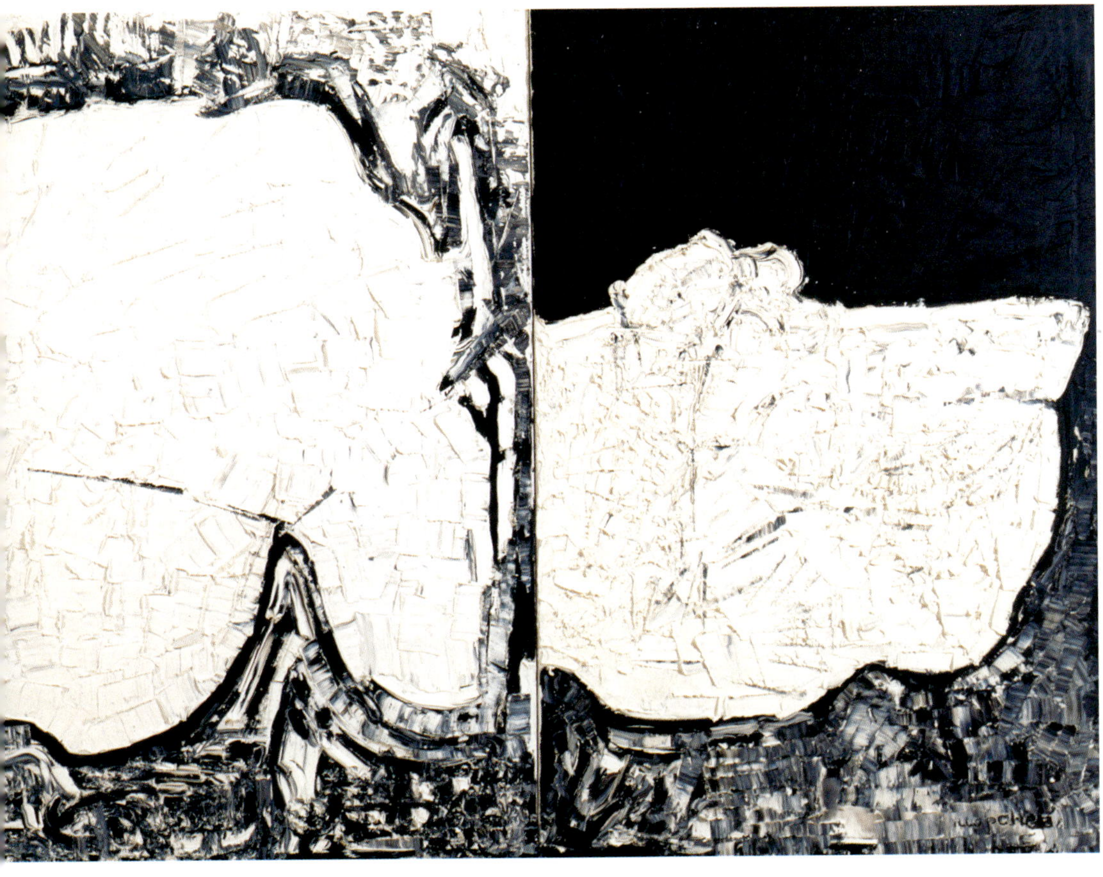

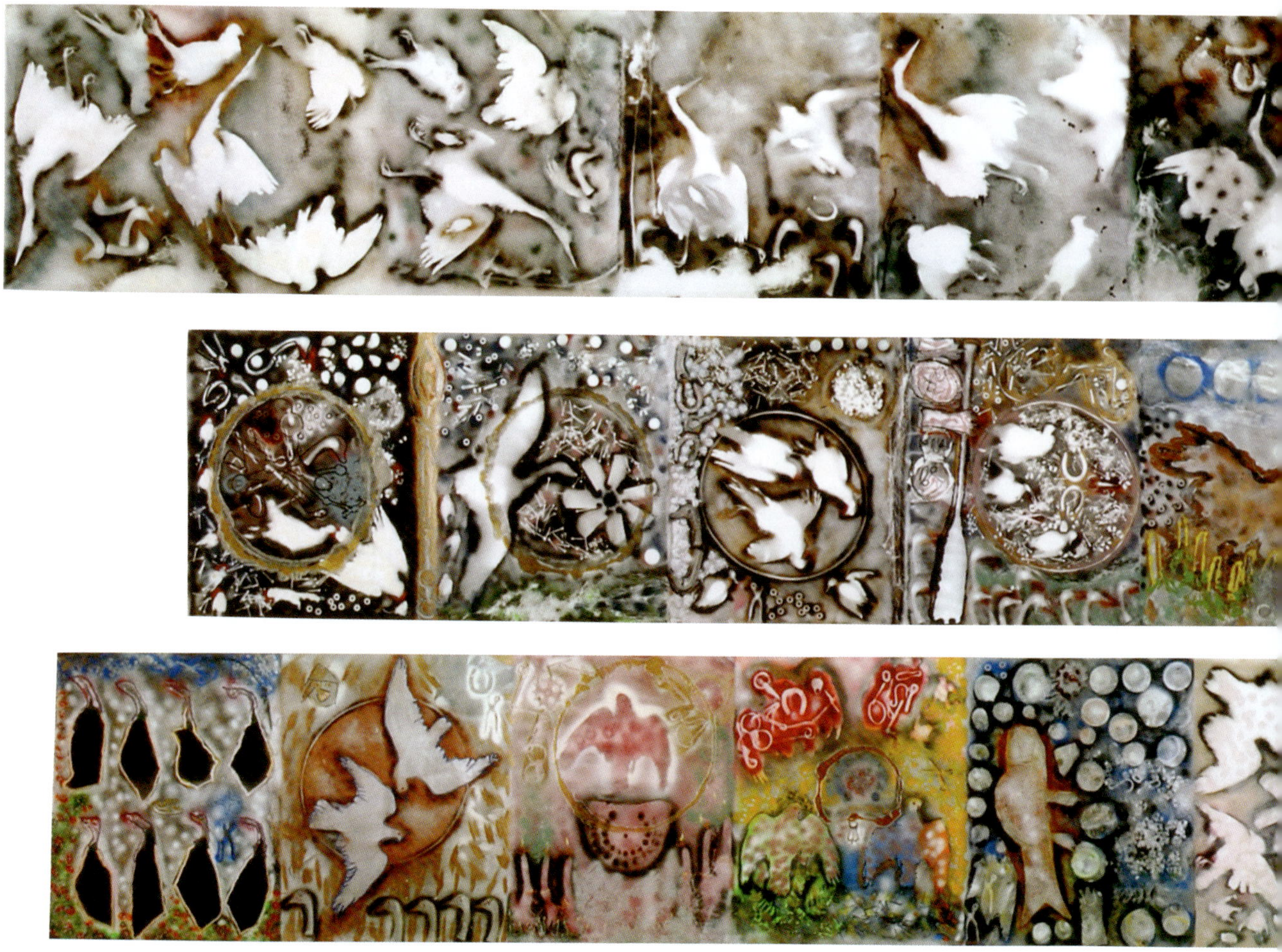

FIGURE 10
Jean Paul Riopelle, *L'Hommage à Rosa Luxemburg*, 1992, acrylic and aerosol paint on canvas, triptych, 155 × 4039 cm (3 canvases; from top to bottom: 155 × 1424; 155 × 1368; 155 × 1247 cm). Musée national des beaux-arts du Québec, accession no. 96.96. Gift of the artist. Catalogue raisonné no. 1992.013.1992.

out was easy. From then on he worked there or in his studio in Saint-Cyr-en-Arthies.

On January 1, 1990, Riopelle returned once and for all to Canada.[15] He set up a studio on Île-aux-Oies, an island that is part of an archipelago in the St. Lawrence River, north of Quebec City. After his first visit there in 1976, he had often returned to hunt or simply to observe the geese that used the place as migratory stopover. He loved the island for its austere beauty and for the sense of connection it afforded with the larger rhythms of nature. "I'm here for the geese," he once said. "Like them, I belong to the island. To have a drink at the 'Bateau ivre,' the restaurant set up on a wreck on the other side, on the headland of Ile aux Grues, and to watch the sun sink into the Saint Lawrence . . ." (Érouart 1994, 23).

For the next two years, dividing his studio time between Sainte-Marguerite-du-Lac-Masson and Île-aux-Oies, Riopelle painted at a fast pace, producing hundreds of works. He painted mixed-media works on wooden panels and paper, using aerosol paint and paint for crafts and fabrics, reworked prints, and made new prints. He also painted (both inside and out) a twenty-eight-foot canoe, of the type used for racing on ice on the St. Lawrence River.

In November 1992 Riopelle learned of the death of Joan Mitchell. In the days following, he produced *L'Hommage à Rosa Luxemburg*, a triptych made of three fifteen-meter-long canvases (fig. 10). This monumental work, a frieze comprising thirty distinct vignettes, is considered Riopelle's final artistic statement. It has been said to convey an

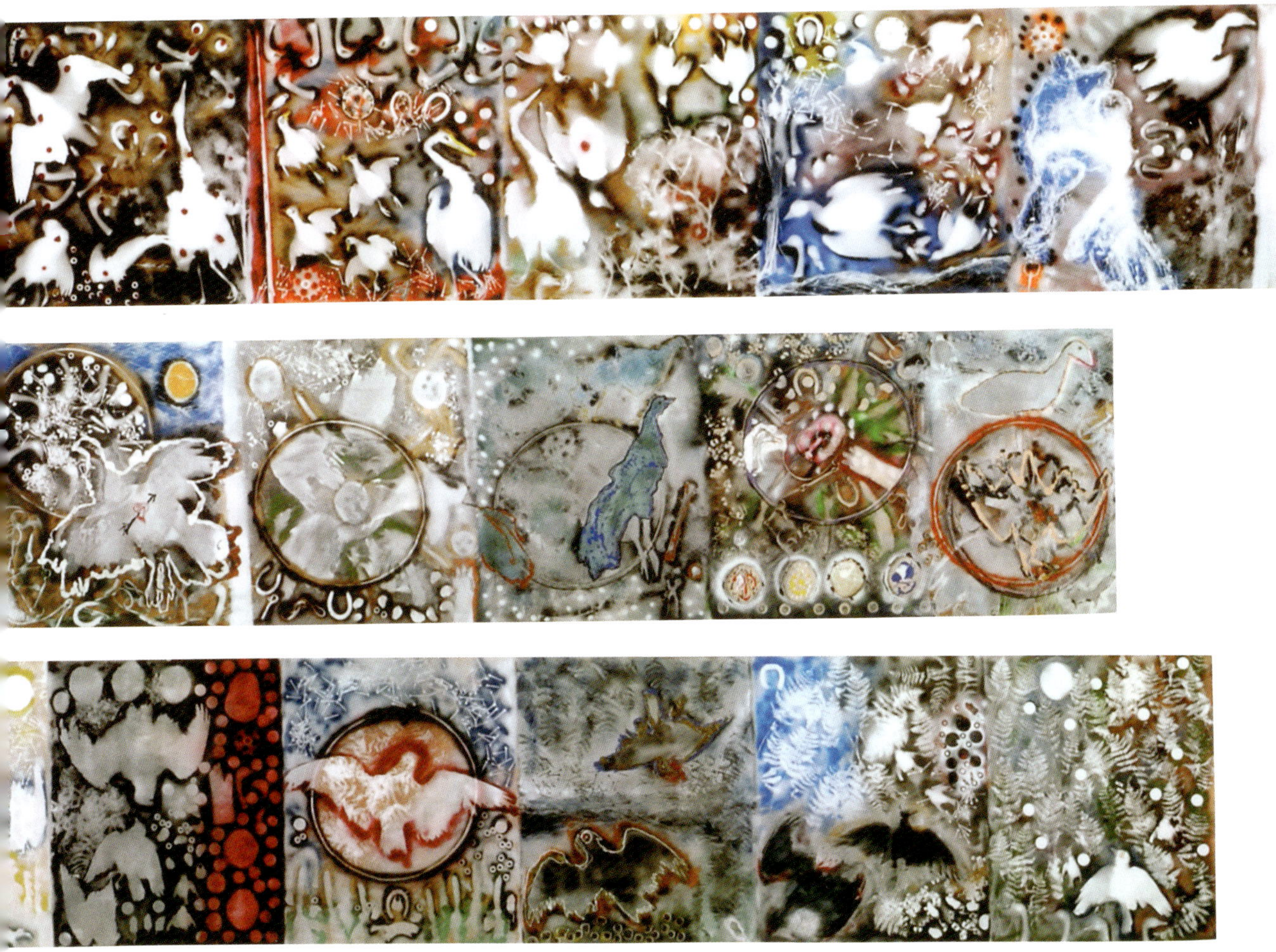

intimate message using objects as words, a secret code that one needs to decipher to understand its autobiographical narrative (Gagnon 2000). Included in a major exhibition held in 1993 at the Michel Tétreault Art International Gallery, it was acquired by the Musée du Québec (now the Musée national des beaux-arts du Québec) in 1996. It would be Riopelle's last painting.

He would, however, remain very much present on the art scene. His works continued to be exhibited in Canada, the United States, and France and elsewhere in Europe. His works also continued to be printed; many silkscreens and etchings were produced, as well as albums that included etchings or lithographs. Two books on his life and career were published in the late 1990s (de Billy 1996; Bernier 1997), and in 1999 the first volume of the catalogue raisonné of his œuvre appeared (Riopelle 1999). The catalogue raisonné—a formidable task, considering his prolific career—was undertaken in about 1985 and will include nine volumes, four of which have been published as of today.

In 1995 Riopelle acquired McPherson Manor on Île-aux-Grues, an island adjacent to Île-aux-Oies. He lived there for long periods, returning occasionally to Sainte-Marguerite-du-Lac-Masson. Riopelle died on Île-aux-Grues on March 12, 2002. The following spring, the Montreal Museum of Fine Arts held a major retrospective exhibition in his honor. His works continue to be exhibited in public and private galleries around the world.

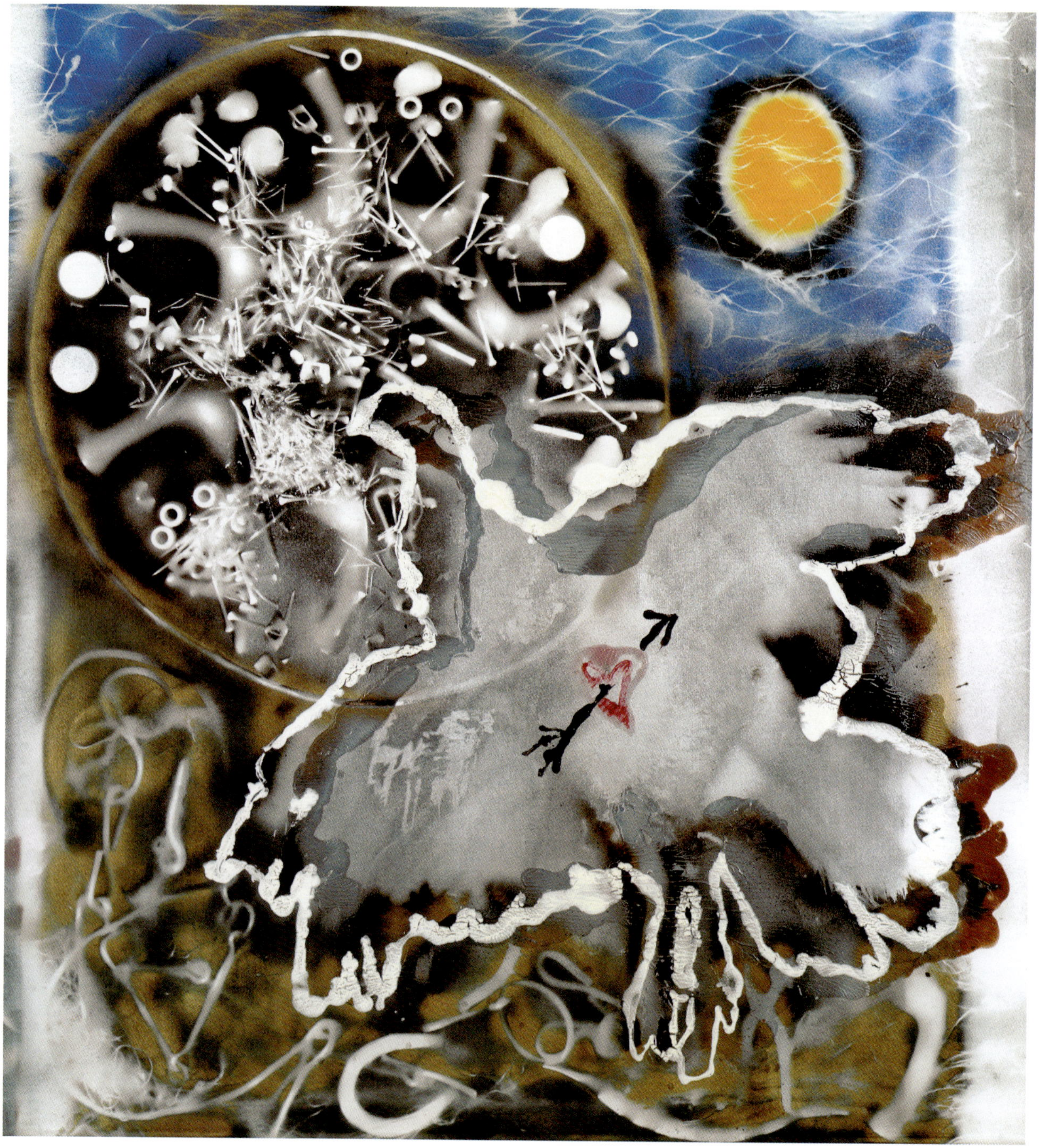

Details of Jean Paul Riopelle, *L'Hommage à Rosa Luxemburg*, 1992, Part 2, panel 6; Part 3, panel 10.

Photographer: Patrick Altman

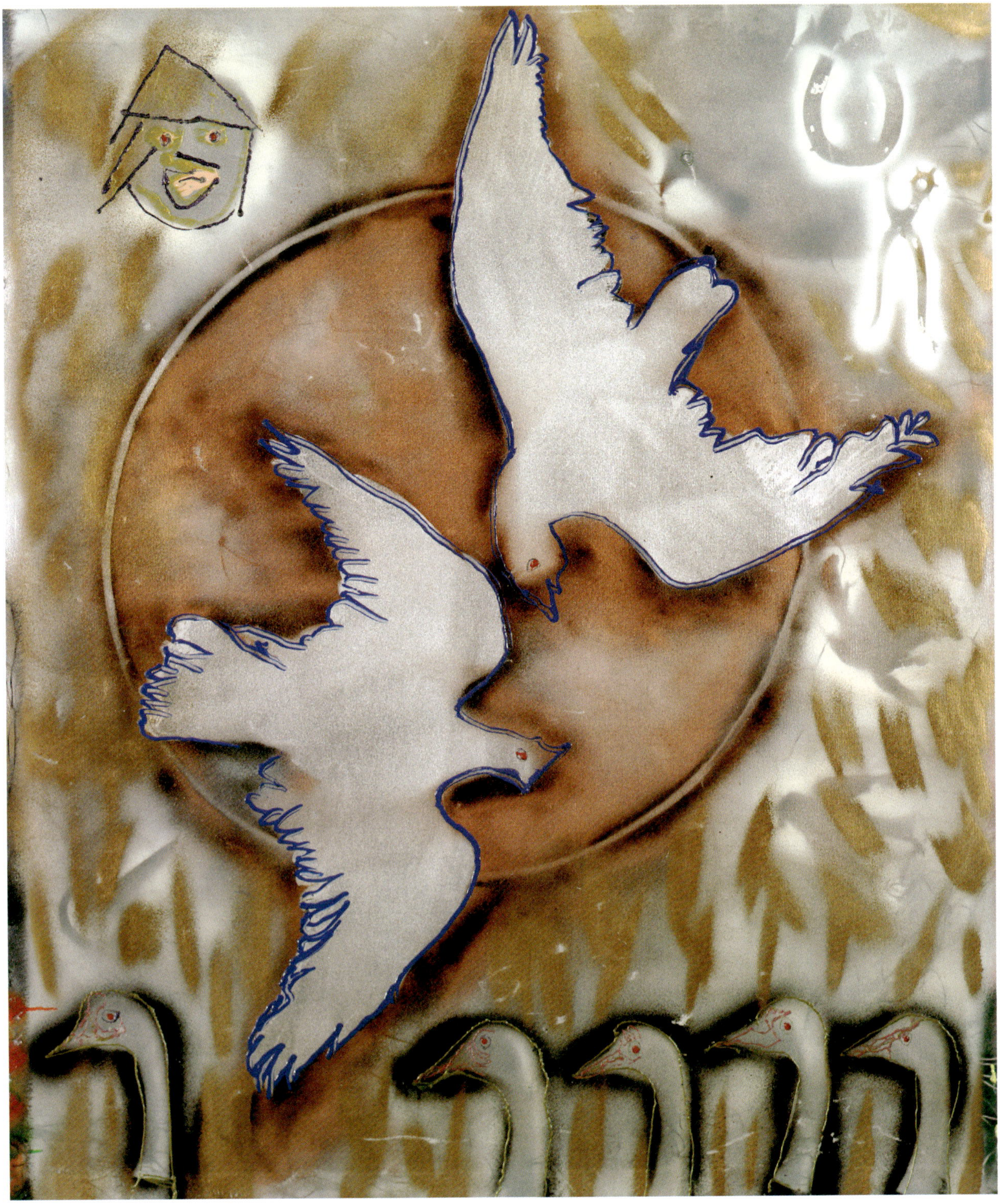

2 Modernist Migrant

Stéphane Aquin

Jean Paul Riopelle's contribution to art history is complex, in part because his œuvre remains undervalued on either side of the Atlantic—both in Paris, where he settled in 1948, and in New York City, where he began exhibiting in 1953. Jane Livingston, in her 2002 catalogue of the retrospective exhibition dedicated to the work of Joan Mitchell—Riopelle's companion for almost twenty-five years—at the Whitney Museum in New York City, noted that "Riopelle's talent was prodigious but is to this day unappreciated in the United States" (Livingston 2002, 25). The art historian Yves Michaud, author of many studies and exhibitions on the art of the 1950s and 1960s, wrote in 1994 that Riopelle was "one of the last major artists of his generation to have been denied the credit he deserves" (Michaud 1994).[16] With the exception of those who knew him in France and have fond and admiring memories of his dazzling personality and unclassifiable œuvre, as well as scattered aficionados in the United States, few people outside of Canada are familiar with Riopelle's name and œuvre today. The Montreal Museum of Fine Arts organized an exhibition that traveled to the State Hermitage Museum in St. Petersburg and the Musée Cantini in Marseille, but otherwise no major exhibition in Europe has been dedicated to his works since the 1981 retrospective organized by the Musée national d'art moderne in Paris. In the United States, despite the featuring of his works in some major collections, including the Hirschhorn and the Guggenheim, the Los Angeles County Museum of Art, and several others, no museum exhibition has focused on his œuvre.

In contrast to his lack of presence on the international scene, Jean Paul Riopelle is seen in Canada as a national treasure, particularly in his native Quebec. There, he has consistently been celebrated by institutions and the public. Tributes, exhibitions, essays and biographies, record sales—everything confirms Jean Paul Riopelle's status as the greatest Canadian modern artist. "C'EST NOTRE PICASSO" (This Is Our Picasso), declared the Montreal newspaper *Le Devoir* the day after his death in March 2002.

The reasons for Jean Paul Riopelle's "regionalization," as the art critic Michel Waldberg termed it (Waldberg 2004, 8), stem not only from the artist's personality—he was notably resistant to stylistic interpretation and to associating with people who could advance his career—and from his personal journey, which finally left him on an island in the St. Lawrence River, far from city rumblings, but also from contextual issues specific to the critics' circles in which his œuvre was received. As a consequence, the profound originality of Jean Paul Riopelle's œuvre and the wealth of his contribution to postwar modern art history remain to be fully appreciated on the American and international scenes.

Montreal, the Automatist Moment

As an up-and-coming young figurative painter enamored by the Canadian landscape of his youth, Riopelle discovered modern art through Paul-Émile Borduas at the École du meuble de Montréal in 1943. Largely unknown abroad, Borduas is recognized in Canada as a kind of father of abstraction and in Quebec as a prophet of modernism. Borduas's *Refus global* manifesto of 1948, signed by Riopelle and a dozen other artists, created a public scandal and resulted in his being fired from the École du meuble. Borduas first retreated to New York City and then to Paris, where he died in 1960, ignored by Parisian circles and forgotten by his compatriots. One gesture of recognition came that year from the Stedelijk Museum in Amsterdam, which, under the direction of the great Willem Sandberg, devoted a posthumous retrospective exhibition to Borduas's œuvre. After that came only silence from abroad. But back in 1942, in the isolation of a Canadian province dominated by the clergy's conservative power, Borduas had developed a clearly original understanding of the Surrealist concept of automatic writing described in André Breton's first *Manifeste du surréalisme*, published in 1924. Borduas interpreted the ideas from a visual rather than literary point of view. In contrast to the Automatism of Parisian Surrealist painters, which consisted of a figurative association of heterogeneous images, Borduas's Automatism required the freeing of subconscious creative forces, his *dictée de l'inconscient*, shunning all figurative or literary mediation. Gestural abstraction was the inevitable result of such a plan. More than just material art, Borduas's Automatist revolution—and it should be emphasized that his was the first systematic abstract reading of Automatism; the painter André Masson's essays in the late 1920s had been an isolated foreshadowing—drew on man's every dimension as it allowed liberation from all forms of culture.

Gradually a group of young artists rallied to Borduas, convinced of the revolutionary social and artistic power of his ideas. Among them was Riopelle, who played a key role not only through his œuvre but also through the ties he developed in Surrealist circles when he traveled to Paris in 1946 and 1947 before settling there permanently in 1948. The young painter began to experiment with decalcomania, a process drawing on chance rather than composition, favored by Surrealists and especially by the Spanish painter Oscar Dominguez. It consisted in folding over, either partially or entirely, a sheet coated with paint so as to produce a set of splotches that were considered very revealing. More important, Riopelle produced a quantity of small watercolors for which he established a vocabulary that would become his own. As the admirable watercolor *Sans titre* (fig. 11) illustrates, he first applied vivid, intense color in large, irregular swaths. Over this colored background he threw a snare of halting, "automatic" black ink lines that had no connection to the background, in this way simultaneously liberating the line, formerly trapped by its function of form delineation, and color, freed at last from all limitations but the frame. These modest-sized watercolors were among the greatest successes of Montreal's burgeoning Automatist movement and brought Riopelle the first words of praise from the Parisian circles he frequented in 1946. The influential art dealer Pierre Loeb, to whom Riopelle showed the works in 1947, wrote a letter to introduce the artist to André Breton, known as the pope of Surrealism. Breton acquired two of the watercolors and kept them in his collection until his death.

For a time, Riopelle participated actively in the conversation between Borduas and his disciples and André Breton, from whom Riopelle and some of his fellow artists received an invitation to contribute to the sixth *Exposition internationale du surréalisme* held at Galerie Maeght in the summer of 1947. Borduas declined the invitation, prob-

FIGURE 11
Jean Paul Riopelle, *Sans titre*, 1946–47, watercolor and black ink on paper, 16 × 23.7 cm. Montreal Museum of Fine Arts, inv. 2001.161. Gift of Jean Paul Riopelle. Catalogue raisonné no. 1946.014P.1946-1947.

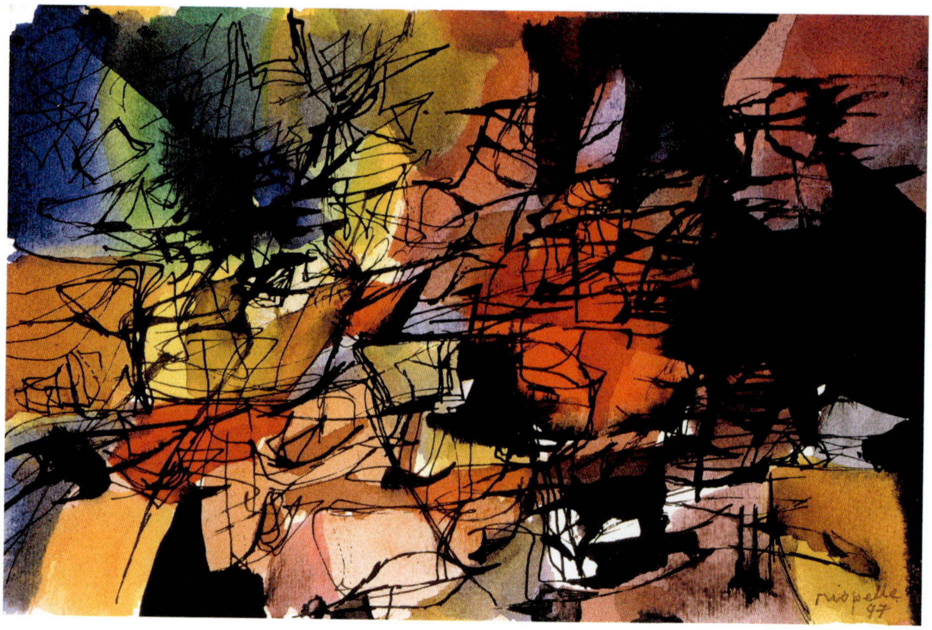

ably fearing that his group would become subordinate to Breton's group, but Riopelle accepted, though he kept his distance from Surrealist orthodoxy. "I am participating as a foreigner," he told a Montreal reporter. "I am particularly a foreigner when it comes to the 'monolithic trends' that France currently favors" (Gagnon 1998, 364). The titles of the two works he showed there, *Testicules de la planète* and *Poil de Dieu*, are apparent proof of his own spirit of insubordination. Earlier that year, he and Fernand Leduc, another Canadian painter who had recently settled in Paris, organized the *Automatisme* exhibition at the newly opened Galerie du Luxembourg, in which the works of the Montreal group were featured.

Although Riopelle's promotion of Montreal's Automatist movement in Paris did not bring international recognition of this movement's originality, it did enhance his stature on the Parisian scene and notably established his independence from any type of organized group or coercive pictorial ideology, a position he never abandoned. From early in his career, Riopelle's freedom of spirit drove his openness to renewal, resulting in a unique adventure in the visual arts. Paradoxically, other than in the initial Automatist phase, his œuvre also suffered from a lack of association with any specific movement in art history. And as Yves Michaud so insightfully noted, "The problem for us and for art historians who need to pigeonhole him (more than for Riopelle himself, whom it in no way prevented from taking his work where he wanted it to go) is that the artist did not seek to defend his development. It fell within the logic of his life as a painter, not under some theory" (Michaud 2009, 51).

The Parisian Years

Though his name is often linked with the so-called École de Paris—the abstract painting scene that flourished in early 1950s Paris, actually a constellation of practices and trends rather than an organized movement—Riopelle remained a foreigner within those circles. As such, he felt unconstrained, and the instinct that seemed to guide him in his adventure in the visual arts was seen as the mark of a kind of primitive energy that was foreign to Europe and, because it opened the way to renewing abstract art's traditional foundations, even more desirable.

All the leading critics in Paris at the time, from André Breton to Michel Tapié, Georges Duthuit, Patrick Waldberg, and Pierre Schneider, became enamored of Riopelle, whose reputation quickly grew in stature thanks to his pictorial exploits and great charisma. Riopelle was not outspoken, but he painted with an energy and a feeling for surface and color that few or none in Paris could match. Much was written on his Canadianism, with some writers trotting out well-worn metaphors about Canadian forests and wide-open spaces. André Breton, in the preface he penned with his wife, Élisa, and Benjamin Péret in the catalogue of the Riopelle exhibition at Galerie Nina Dausset in 1949, described Riopelle as a "master trapper," a concept that would become inseparable from the artist's image. Visually speaking, Breton was exactly right. "To me, this is the art of a master trapper," he wrote. "Traps are laid both for animals in burrows and those in the air, as Germain Nouveau said. What wins me over to the idea of a trap, which I guardedly accept, is that they are also traps for traps. Once these traps are trapped, a great degree of freedom can be attained" (Breton, Breton, and Péret 1949). The works to which he referred are indeed characterized by gossamer networks of lines and colors that "trap" each other in a space where Riopelle allowed himself all gestures and thrusts, runoffs, scratches, knife strokes, splotches, and spatter.

In 1951 the influential critic Michel Tapié included Riopelle's works in a landmark exhibition, *Véhémences confrontées*, presented first in Paris and then in London the following year. It was the first major postwar world event in which abstract painters of various inclinations participated. Next to Riopelle were Jackson Pollock, Willem de Kooning, Giuseppe Capogrossi, Camille Bryen, Hans Hartung, Wols, and Georges Mathieu. It should be noted that the early 1950s Paris scene was not as French as one might have thought but was instead an international platform—one that today would be called global. Tapié, himself a great traveler and later an influential figure for the Japanese Gutai group that he promoted in Europe, provided a crucial example in defining Paris as a crossroads for the most radical advances in abstract art. But there was also Georges Duthuit—Matisse's son-in-law, a specialist in Byzantine art and, like Breton, a great aficionado of Canada's West Coast masks—who saw Riopelle and his Paris-based friend Sam Francis as protagonists of a renewal of European tradition with the influx of a more "instinctual" North American culture, which could result in repositioning Paris as an artistic metropolis.[17]

From the first sentence of his preface to the first Riopelle exhibition at the Pierre Matisse Gallery in New York City in 1954—translated from French by Samuel Beckett, a great friend of the artist—Duthuit provides an eloquent explanation of that opinion: "Like a trapper fresh from the Canadian solitudes measuring his stride to our narrow pavements, Jean-Paul Riopelle seems hardly to contain the flooding energies of youth at its full" (Duthuit 1954). Duthuit developed a rich analysis of Riopelle's work from that somewhat folkloric vision, calling to mind Harold Rosenberg's contemporary reading of Action Painting. His analysis is worth quoting at length:

> What a stupendous exertion of strength is involved in this stripping of oneself bare of all that is not pure intuition! Tons of colour employed to prove, layer upon layer, spasm after spasm, that the spirit is rid of connectedness (a miracle, incidentally, that the outcome should be transparence and delicacy). He must be constantly on the watch lest something of conscious logic steal into the picture, which the least of its incursions would render null and void. The work must be completed in a single stretch—sometimes lasting ten hours—in a

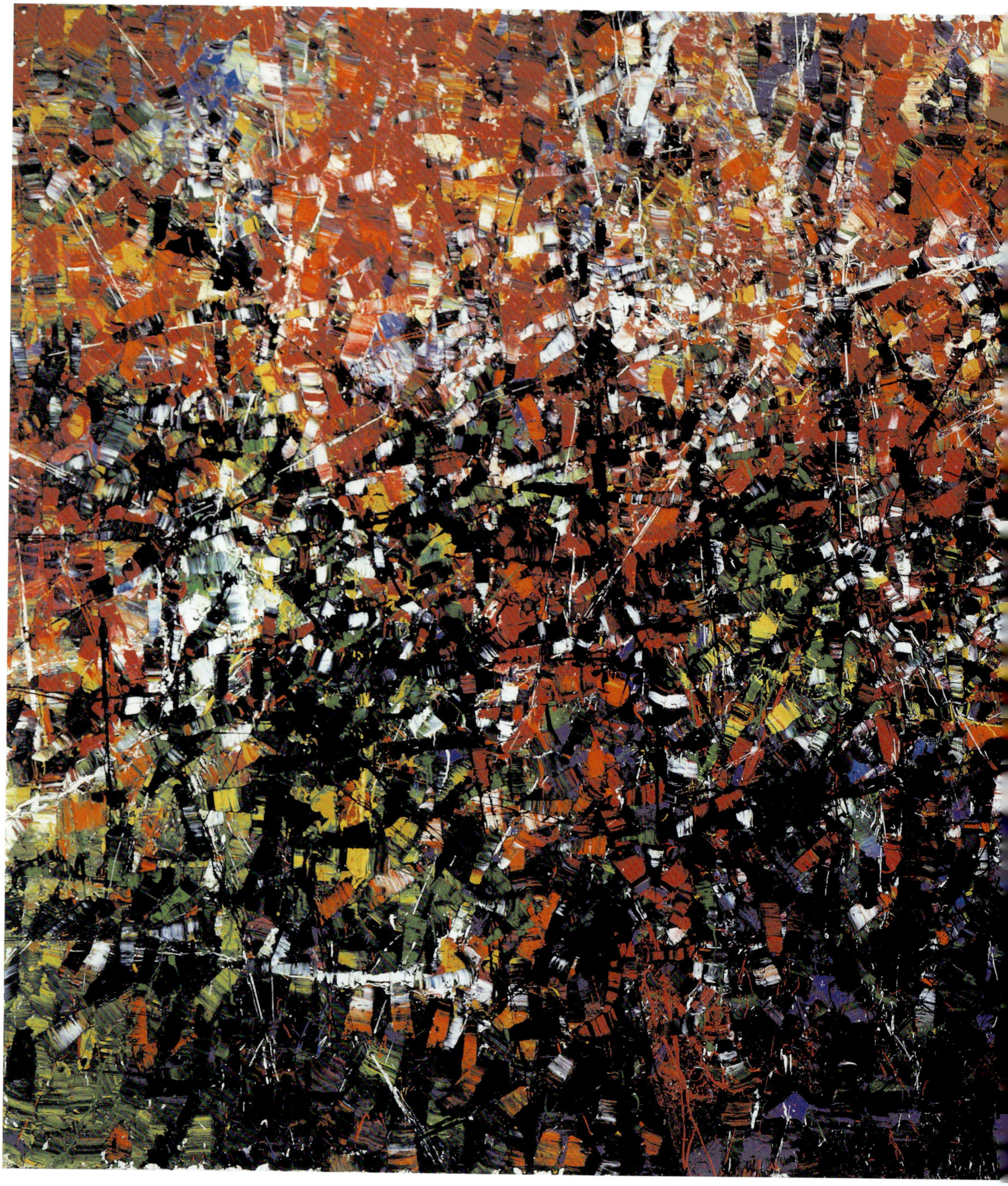

FIGURE 12

Jean Paul Riopelle, *Sans titre*, 1953, oil on canvas, 198.2 × 350.5 cm. The Power Corporation of Canada Art Collection, inv. 1993.7.1. Catalogue raisonné no. 1953.055H.1953.

Photo: The Power Corporation of Canada Art Collection

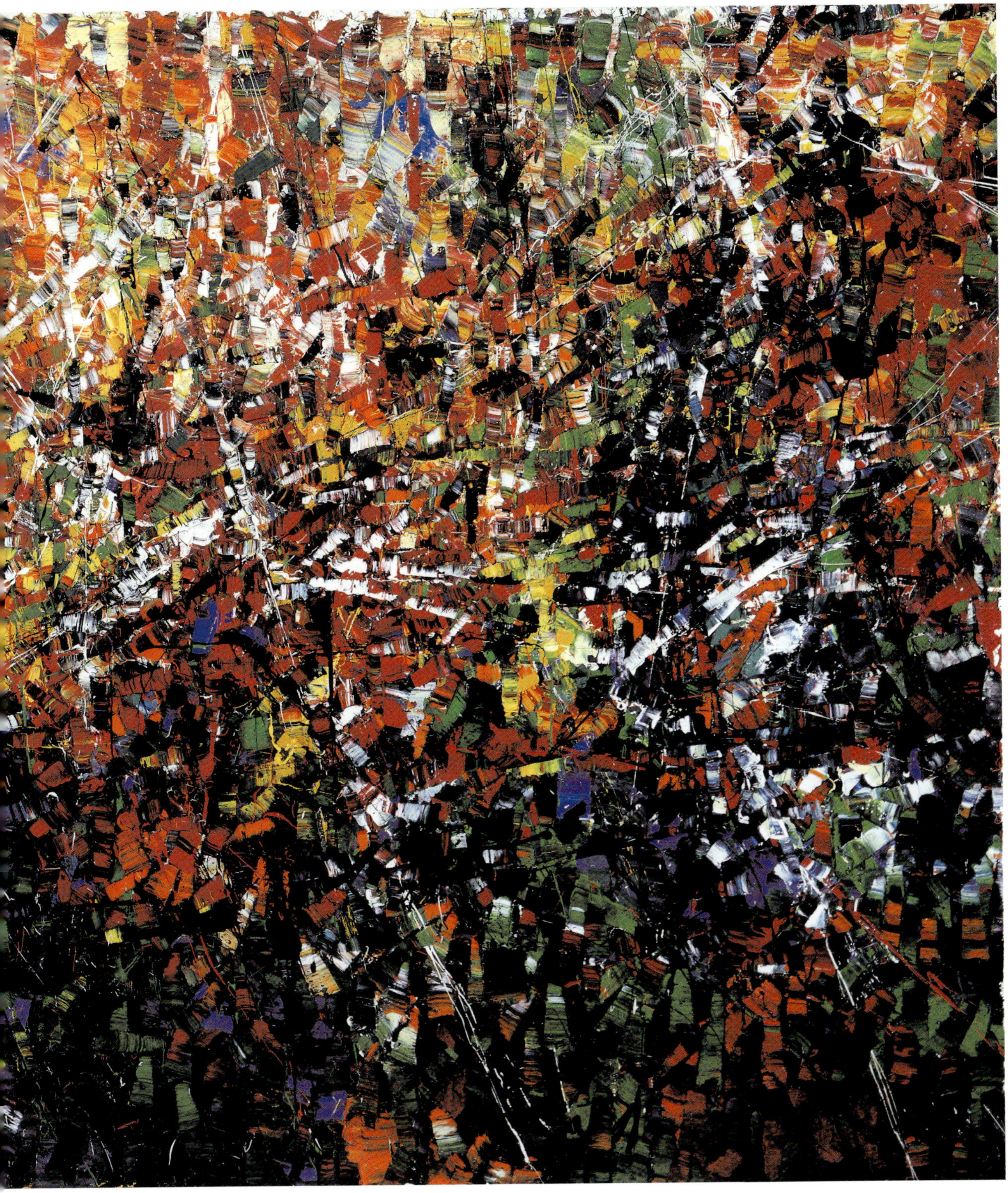

kind of vertiginous rapture, invested from all sides at once and with no other guide than the feeling that somehow or other it will be accomplished. And the picture is completed when at last there appears in its glistening nudity the sole emotional flux whence its undertaking proceeds and by which now its aspect is determined. (Duthuit 1954)

As he wrote that text, which he began in 1951 and finished in 1953, Duthuit could observe the first "mosaics," extraordinary masonries of color for which Riopelle is famous (fig. 12). From edge to edge, the entire surface is covered with a marquetry of thick and dense, energetic and nervous strokes that give the works a rare dynamism. Straight out of the tube, color is applied with a knife directly to the canvas and sometimes as a spatter of paint. With its all-over-ness, it covers the entire canvas without hierarchy, but the painted surface is far from flat. There is no flatness here. Textured, dense, deep, and uneven, it presents itself as a wall of paint, a mosaic of pure color.

American Reviews

While French critics praised Riopelle's "American-ness," Americans criticized his European sophistication. New York City discovered Riopelle in 1953 when the Guggenheim Museum held the *Younger European Painters* exhibition, organized by James Johnson Sweeney. The context was not ideal. Contrary to what was happening in Paris, a climate of nationalism prevailed in America; New York art circles, busy with "stealing the idea of modern art"[18] away from Europe, sought to establish the superiority of local abstraction over that of the old countries. In response to a study by *Art Digest*, published a few months before the Guggenheim exhibition, under the heading "Symposium: Is the French Avant-Garde Overrated?" Clement Greenberg wrote, "Do I mean that the new American abstract painting is superior on the whole to the French? I do" (Greenberg 1953, 12). He elaborated: "Our new abstract painting seems to have anticipated the Paris version by two or three years, but I doubt whether there has been, or is yet, any real acceptance of American influence on the part of the French (and I don't much care)" (Greenberg 1953, 27). Amply commented on by the press and in art circles, the exhibition was quickly placed at the heart of this debate. Riopelle, who all agreed was one of the best of the group, was caught in the crossfire of such unfortunate reviews. The critic Otis Cage deplored European painters' lack of "painterly sensitivity" and qualified Riopelle's art as "boring": "The fresh impulses, the new directions, are on this side of the water. The younger Europeans realize that, even if Sweeney does not" (Cage 1953).

Even though the Guggenheim acquired a Riopelle canvas (*La Nuit bleue*, 1953) and the Pierre Matisse Gallery put the artist under contract, Riopelle was never really able to shed his undesirable affiliation with the École de Paris. Worse, he was seen as adopting a mannered style of American Abstract Expressionism, notably that of Jackson Pollock. In 1954, when Riopelle first exhibited at the Matisse Gallery, the critic Sam Hunter compared him to Pollock, saying that he was less "inspirational" and deploring his "fussiness" and the "over-emphasis on virtuoso paint manipulation," which were all, in the end, European traits (Hunter 1954).

Fifty years later, Riopelle was still being tarred with the brush of those first critical reviews. Commenting on an exhibition of Riopelle's works at the Robert Miller Gallery in 2005, Ken Johnson began his *New York Times* article with a reference to Greenberg's early assessment: "In a short 1953 essay, critic Clement Greenberg answered yes to his question 'Is the French Avant-Garde Overrated?' He did not mention Jean-Paul

Riopelle by name, but judging by this selection of paintings from the 50's and 60's it appears that the Paris-based Canadian was one of the artists Greenberg had in mind when he criticized the French for creating a tame and prettified version of Abstract Expressionism" (Johnson 2005).

Johnson's comment shows only too well the general lack of knowledge about the radically original Automatist sources of Riopelle's work, which owe nothing to American Abstract Expressionism, as well as the complexity of artistic issues in postwar Paris. Also ignored, here as elsewhere, was the crucial role Riopelle played in developing the work of a small number of American painters who settled in Paris after the war, including Norman Bluhm, Sam Francis, and, of course, Joan Mitchell, the artist's companion from 1955 to 1979. Though barely older than these other painters, Riopelle had considerable influence on them through his painting, knowledge, charisma, and connections. It is he, for instance, who introduced Francis to Georges Duthuit and Michel Tapié, two encounters that proved crucial in the young American's artistic development. Technically, Riopelle's fluid, all-over use of one gesture was taken up by Francis in his work of 1952 to 1954. His recurrent use of the triptych from as early as 1954—a novelty in the field of Abstract Expressionist painting—came to be a regular compositional device in Joan Mitchell's art. Riopelle also succeeded in sharing his love of Monet with his companion, who settled in Vétheuil, one of Monet's residences, and whose painting displayed an ever-increasing fluidity. The breadth of the artistic collaboration between the lovers is still to be assessed. In a rare essay on their relationship, the art historian David Moos wrote, "Despite copious biographical and art world intersections between the itineraries of the two painters, it is remarkable that in most scholarly studies of Riopelle and Mitchell there is little and in many cases no significant mention of the other" (Moos 2004). Those involved, however, recognized that influence. Recalling the relationship, Norman Bluhm said: "She [Mitchell] drew much from him.... However, Americans do not recognize the influence Riopelle had on Joan because Jean-Paul's work is ignored in the U.S. For my part, I knew them intimately, and I can say that she owes him a lot" (de Billy 1996, 127). It can be said with certainty that some part of the history of American abstract painting, through the work of these illustrious expatriates, owes Riopelle its absorption of Matisse's hedonism, of Monet's fluid sense of color, and of Courbet's brutal materialism.

A World of His Own

Art history tends to reduce the work of artists to that moment when their path meets a particular chapter in the great narrative of the evolution of artistic trends. For Riopelle, that moment was postwar Paris Abstraction, complete with the vagaries of critical reviews. As is the case for any really great artist, Riopelle's work has its own logic; it may be anchored in or associated with a particular moment of history, but it remains independent of trends and fashion and, over the course of several decades, was injected with an authentic force of renewal and transformation. After the 1950s, Riopelle's work continued to evolve in an implacably original way, with a commendable modern focus on breach of convention and innovation, regardless of success or standards. To a question by Gilbert Érouart regarding the painters who moved him, the often unforthcoming Riopelle responded with words that applied to himself perhaps more than they did to anyone else: "I've always been moved by those who were able to transcend the system that constricted them" (Érouart 1995, 10).

Research has only begun to shed light on the succession of pressure points to

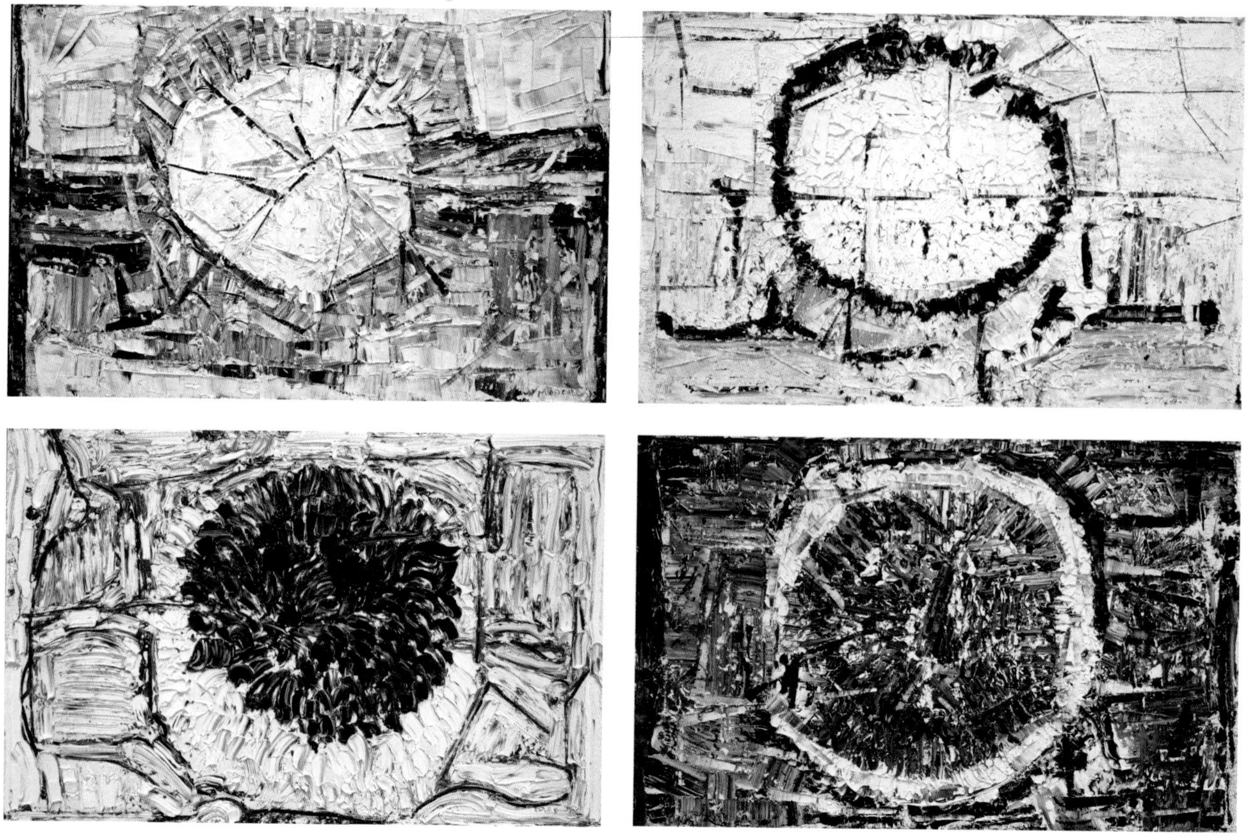

FIGURE 13
Jean Paul Riopelle, *Soleil de minuit (Quatuor en blanc)*, 1977, oil on canvas, quadriptych (4 canvases, 97 x 146 cm each). Montreal Museum of Fine Arts, inv. 2001.164.1-4. Purchase with a special grant from the Government of Quebec. Catalogue raisonné no. 1977.008H.1977.

FIGURE 14
Jean Paul Riopelle, *Mi-carême* (recto of *Merry Go Round*), 1990, mixed media on wood panels, 203 × 131.7 cm. Montreal Museum of Fine Arts, inv. 2001.160.1. Gift of Jean Paul Riopelle. Catalogue raisonné no. 1990.236.1990.

which Riopelle's artistic trajectory was subjected from the 1950s until his death. The pictorial maturity of the 1960s—which earned him a UNESCO prize at the 1962 Venice Biennale, where he represented Canada—is lush with monumental works. The freedom with which Riopelle applied color in these works ushered in a period of unprecedented hedonism in painting. Often organized in diptychs and triptychs, his paintings of the period flaunt an astonishing indulgence in fleshy paint, balanced by a compositional awareness that shows him moving away from the instinctual all-overness of his early years. Distinct forms of contrasting colors make their way through an ever-heavier handling of paint. Nothing could be further from Clement Greenberg's ideal of the time—flat, monochromatic color fields—than Riopelle's painting of the 1960s.

Along with his painting, Riopelle began sculpting during that decade. His œuvre in that field is among the most remarkable of the period, rivaling that of Willem de Kooning, with whom he shared a sense of shape and texture directly inspired by Rodin's excess and by his great personal friend Alexander Calder's carnivalesque vision. First produced in plaster in 1969–70 and presented to the Maeght Foundation in Saint-Paul-de-Vence in 1970, the monumental *La Joute* was cast in bronze a few years later. It is a vast ensemble of characters—among them an Indian chief, an owl, and a bear dressed in female attire—given life with a play of water and light features. It is considered one of Montreal's landmarks of public art.

Riopelle also played a significant role in the area of printed art, which he discovered in 1966 when he joined Galerie Maeght in Paris. Founded in 1947 by Aimé Maeght, Galerie Maeght became one of the most prestigious European galleries of the early 1960s and represented some of the greatest modern artists of the period, including

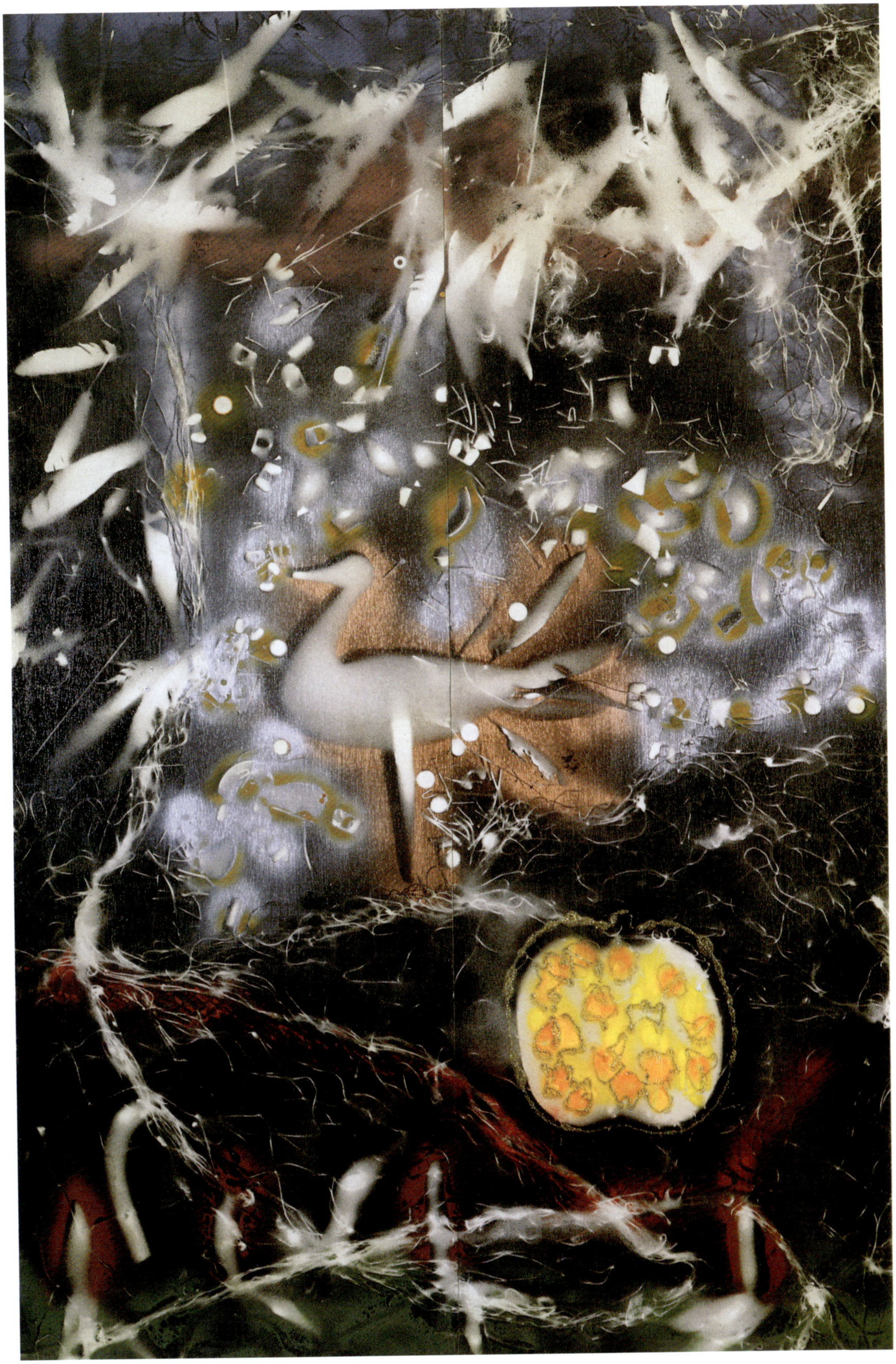

Braque, Giacometti, and Chagall. Thanks to the printing firm Arte and its lithographic and engraving workshops in particular, artists were allowed to experiment in new areas. Riopelle excelled there, freely exploring various printing techniques with great genius[19] and producing more than three hundred original works (Riopelle 2005).

It was in printed art that figurative expression made a comeback in Riopelle's work. These figurative works manifest once again his exemplary avant-gardism. After he began returning more often to his native land, in the 1970s, northern motifs, Inuit string games, polar landscapes, and the migration of white geese populated his works, bringing his compatriots to see the artist as a kind of national high priest. From a material perspective, one innovation followed another. For a brief time he abandoned color in the stunning 1977 *Icebergs* series (fig. 13), inspired by a flight over the Great North while on a fishing trip, with heavy lines built from an accumulation of painted matter marking the stark contrast between light and dark. These surreal and heroic landscapes bring to mind the influences of Courbet, Rouault, and Van Gogh. Prominently displayed in the touring retrospective organized by the Musée national d'art moderne in Paris in 1981, which circulated to Quebec City, Montreal, Mexico City, and finally Caracas, these monumental paintings have still never been seen in the United States.

As his physical strength diminished, Riopelle found it increasingly hard to manipulate palette knifes loaded with paint. In the late 1980s he took up painting with aerosol cans, using all manner of objects as stencils (fig. 14). The technique was not unlike one he had used for some of his earlier Automatist works, when he sprayed colored ink on sheets of paper through a straw. The resulting works demonstrate an impressive formal inventiveness, sense of rhythm, and mastery of scale. In 1992, at the news of Joan Mitchell's death, he undertook his most monumental work ever, *L'Hommage à Rosa Luxemburg*, an epic cascade of brilliant and ghostly images that only silence could follow with adequate grandeur. He would never paint or draw again.

Unfortunately, Riopelle's self-imposed rural isolation limited the influence of his last works. A fair reevaluation of his entire œuvre will no doubt result in a better appreciation of the great value of these final works and redefine the role Riopelle played in the overall history of modern art. In an era of division, Riopelle served as a conduit between European and North American sensibilities as well as between different aesthetic movements such as Surrealism, Abstraction, and the great modern pictorial tradition that ran from Courbet through Monet and Cézanne to Matisse.

3 Master of Oil Paint

While an emerging artist in the 1940s, Riopelle worked in oil, which he used until the late 1970s. Among his most renowned paintings are those in the *Mosaïques* series, dating from the 1950s, in which firmly applied strokes of the palette knife create a juxtaposition of colored elements. After the very dense *Mosaïques*, his palette knife strokes became larger and more fluid. Color still dominated his paintings of the 1960s and 1970s. In his last series of large-scale oil paintings, *Icebergs*, from 1977, black and white prevailed. A list of the works chosen to illustrate Riopelle's painted œuvre is given in table 1, with reference to the page numbers where the images can be found. Table 2 lists the thirty-three oil paintings executed between 1944 and 1977 for which sampling and analysis was undertaken. Details of the analytical techniques used are given in Appendix I.

Riopelle's oil painting technique was heavily based on creating impasto, with volume playing as important a role as color. Paint was often applied in thick layers. Many crests and fine peaks were formed when the palette knife loaded with paint left the surface of the canvas (fig. 15). Riopelle claimed that the thickness of his paint was unintentional. "When I begin a painting," he said, "I always hope to complete it in a few strokes, starting with the first colours I daub down anywhere and anyhow. But it never works, so I add more, and more, without realizing it. I have never wanted to paint thickly, paint tubes are much too expensive. But one way or another, the painting has to be done. When I learn how to paint better, I will paint less thickly."[20]

In addition to color and volume, a third element, range of gloss, plays a crucial role in Riopelle's work. Matte and shiny paints are juxtaposed in his oil paintings, creating

TABLE 1 Paintings illustrating Riopelle's oil technique

Title	Year	Page number
Nature morte au violon	1941	5
Le Perroquet vert	1949	10
Chevreuse II	1953	12
Sans titre	1953	26, 27
Rencontre	1956	12
Abstraction	1958	39
Abstraction	1958	60
Point de rencontre	1963	14
Mitchikanabikong	1975	56–57
Pangnirtung	1977	16–17
Soleil de minuit (Quatuor en blanc)	1977	30

TABLE 2 Oils on canvas included in the study

Title	Year	Dimensions (cm)	Collection*	Acc. No.
Saint-Fabien	1944	30.4 × 40.9	MNBAQ	78.140
Peinture	1947–48	109.5 × 117.5	MMFA	1982.28
Le Perroquet vert	1949	110.2 × 140	MNBAQ	69.36
Sans titre	1949	81.1 × 100.1	MACM	A 92 1164 P 1
Espagne	1951	150 × 232	MNBAQ	2001.02
Composition	1951	194.9 × 129.8	MACM	A 92 446 P 1
Vent traversier	1952	200 × 300	MMFA	227.1992
Crépusculaire	1953	130 × 162.2	Musée Cantini	C 82-28
Chevreuse	1954	301 × 391	FNAC (attributed to MNAM)	AM 3881 P
Autriche	1954	200 × 300.7	MMFA	1963.1395
Poussière de soleil	1954	245.2 × 345.3	MNBAQ	2000.01
Abstraction	1954	129.7 × 161.9	MNBAQ	63.47
La Roue – Cold Dog (Indian Summer)	1954–55	250 × 331	MMFA	2001.163
Le Cirque	1955	89 × 116	MMFA	1956.1156
Feux-follets	1956	54.4 × 45.4	MACM	A 71 126 P 1
Envol	1956	97 × 130	Canada Council for the Arts – Art Bank	ABBA 73/4-0986
L'Érieux	1957	129.5 × 195	MACM	F 75 51 P 1
Abstraction	1958	22 × 33	MAMC	P1991-0360
Chicago II	1958	250 × 300	MNBAQ	2001.154
Landing	1958	200 × 375	MACM	A 68 56 P 1
13 à la mode	1959	128.3 × 193.5	MACM	A 67 21 P1
Autre pôle	1961	59.7 × 72.9	MACM	A 78 34 P 1
Un Coin de pays	1962	200 × 200	Canada Embassy in Paris	n/a
Vertige	1962	97 × 130	MMFA	22.2002
Point de rencontre	1963	428 × 548 (3 vertical canvases of 286 × 182.7 on top of 2 horizontal canvases of 142 × 274)	FNAC (attributed to Paris Bastille Opera House)	90069
Les Masques	1964	195 × 357 (3 canvases, 195 × 199 each)	MNBAQ	2002.25
Sans titre	1967	38 × 45.6	MNBAQ	72.80
Hommage à Grey Owl	1970	299.5 × 400	MMFA	2001.184
Chemin d'hiver	1973	91.5 × 60	MACM	D 92 1361 P 1
Mitchikanabikong	1975	195.5 × 391.5 (3 canvases, 195.5 × 130.5 each)	MNAM	AM 1979-254
La Mi-été chez Georges	1975	5 canvases; from left to right: 49.5 × 49.5; 161.5 × 97; 49.5 × 50; 49.5 × 61; 49.5 × 50	MNAM	AM 1994-289
Soleil de minuit (Quatuor en blanc)	1977	4 canvases, 97 × 146 each	MMFA	2001.164.1-4
Pangnirtung	1977	200 × 560 (3 canvases; from left to right: 195 × 130; 200 × 300; 195 × 130)	MNBAQ	97.113

* FNAC: Fonds national d'art contemporain (France); MACM: Musée d'art contemporain de Montréal (Canada); MAMC: Musée d'art moderne de Céret (France); MMFA: Montreal Museum of Fine Arts (Canada); MNAM: Musée national d'art moderne (France); MNBAQ: Musée national des beaux-arts du Québec (Canada).

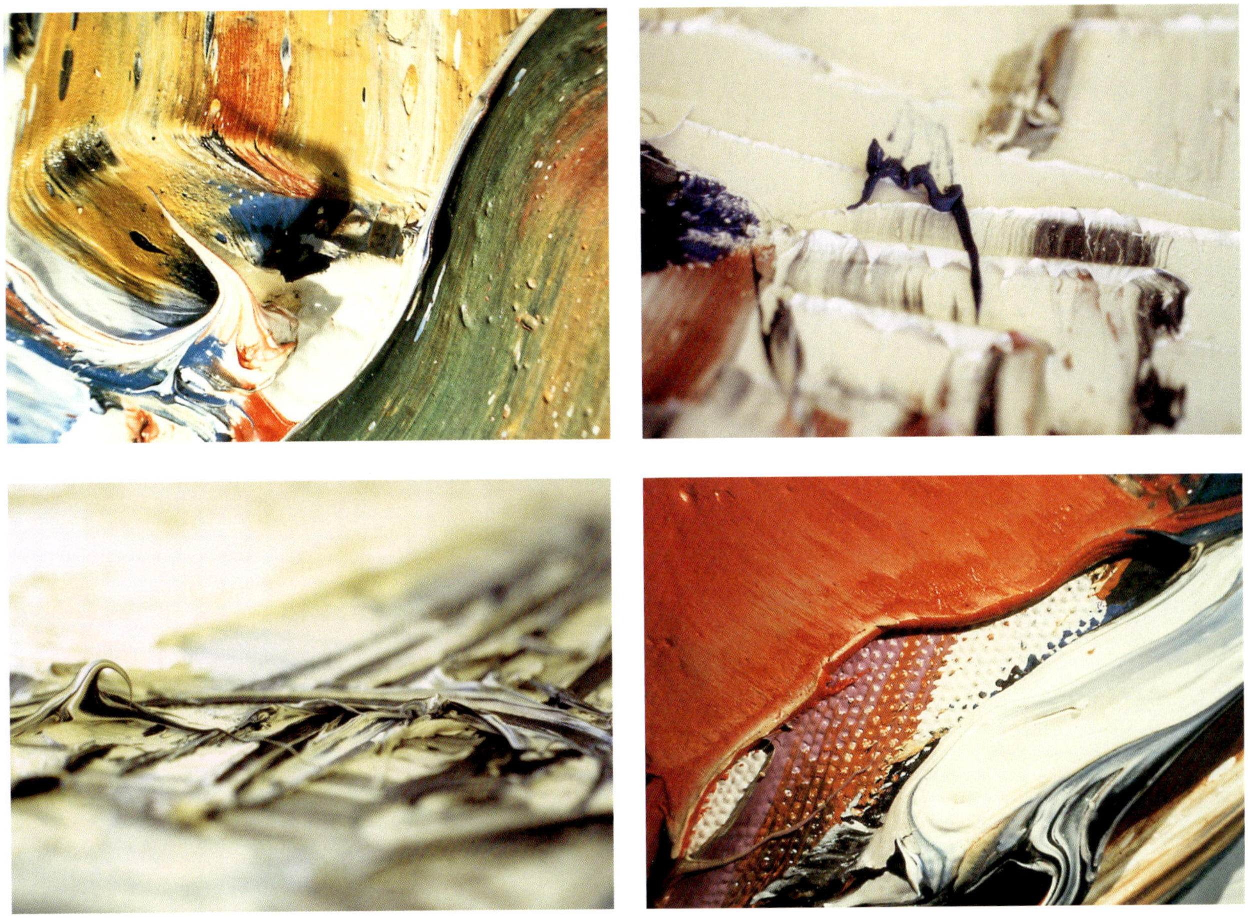

FIGURE 15A–D
Details of *Point de rencontre* showing paint crests, fine peaks, and impasto.

Photos © Canadian Conservation Institute.

areas that reflect light in different ways. Although his style constantly changed during his long and prolific career, his oil painting technique was always based on these three elements—on color, volume, and range of gloss. Jeffrey Spalding has emphasized the preeminence of matter in Riopelle's work: "In Riopelle…the quality of the material substance of paint carries more weight than the image that it is contorted to represent" (Spalding 2002, 13).

The "quality of the material substance of paint" is determined not only by the masterful application of volumes and colors but also by the quality of the materials used, of which large volumes were required to lay down thick layers of paint on the large-scale canvases Riopelle favored. In 1972 Riopelle mentioned during a conversation with the art critic Pierre Schneider, referring to his paintings of the early 1950s, that he used paints that were not expensive (Arbour and Tardif 1972). Ten years later, in a conversation with Jean-Louis Prat, then director of the Maeght Foundation, Riopelle mentioned that he used to make his paints himself during this period; to a somewhat skeptical Prat, Riopelle said that not only did he make his paints but he put the paint in tubes as well, as paint was better protected and kept longer in tubes (Letarte and Feaver 1982).[21] Unfortunately, no details were provided as to where he purchased the raw materials and tubes.

That Riopelle made his paints to save money or used inexpensive paints does not mean that these paints were of poor quality, and paintings from the early 1950s do not show signs of deterioration that could be attributable to paints of lower quality. However, according to the Automatist writer Claude Gauvreau, paintings that Riopelle made in Canada prior to leaving for France in 1948, during a period that could be con-

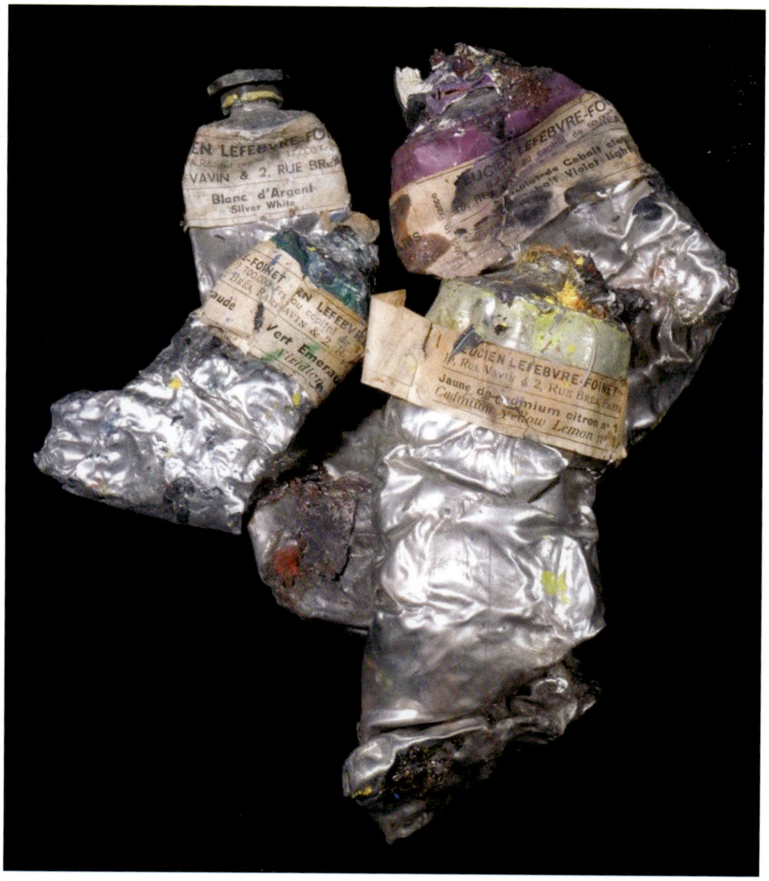

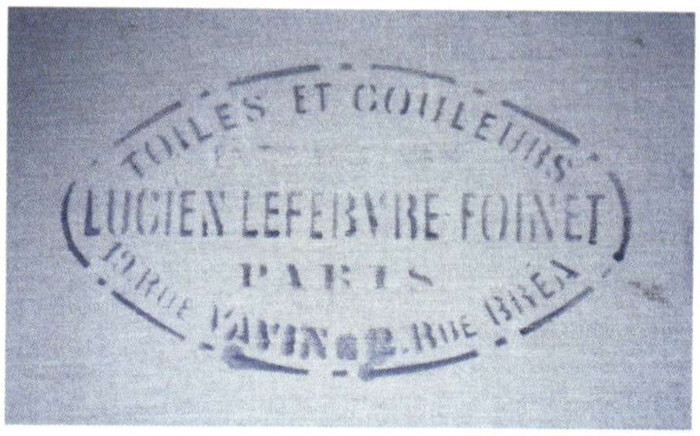

FIGURE 16
Tubes of Lefebvre-Foinet paint recovered from Riopelle's studio.

Photo © Canadian Conservation Institute.

FIGURE 17
Lefebvre-Foinet stamp on the back of the canvas of *Composition*, dated 1951.

Photo © Canadian Conservation Institute.

FIGURE 18
Riopelle in 1953.

Photograph by Denise Colomb. © Denise Colomb-RMN. Photo courtesy of Ministère de la Culture/Médiathèque du Patrimoine, Dist. RMN/Art Resource, NY

sidered an exploratory phase, fell apart because they were made with Sherwin-Williams house paints to save money (Jolicoeur et al. 2002). That may explain why there are very few paintings from this period.

René Detroye has stated that Riopelle made his paints until 1952 and that from about 1952 to 1978 his paints were manufactured by Lucien Lefebvre-Foinet (Spalding 2002, 12). Riopelle's use of Lefebvre-Foinet paint was corroborated by an inventory of paint tubes retrieved from his studio in France after his death (fig. 16). Out of ninety-two tubes, sixty-two are from Lefebvre-Foinet; fifteen others, although their damaged labels make definite identification impossible, resemble the ones from Lefebvre-Foinet. Riopelle also bought his canvases from Lefebvre-Foinet, as shown by the colorman stamp found on the back of the canvas of many of his paintings (fig. 17). Yseult Riopelle, however, has identified paint cans and stretchers in photographs from the early 1950s (fig. 18) as Sennelier products and received confirmation from the supplier that Riopelle indeed bought material from that store for many years; she has also mentioned Édouard Adam (of Adam Montparnasse) as another supplier, and good friend, of her father (Riopelle 2009b). Adam has specified that he sold different types of supplies to Riopelle over the years: stretched canvases, Lefranc-Bourgeois oil paints, and paints prepared by Adam himself, brushes, and other items (Adam 2010). Given the large quantity of paint that Riopelle was using, it would not be surprising for him to have purchased paint from several suppliers.

Lucien Lefebvre-Foinet, a store then located in the 6th arrondissement of Paris, was a family business that supplied materials to artists such as Henri Matisse, Alberto Giacometti, Fernand Léger, and Pierre Soulages, as well as Riopelle and Joan Mitchell.

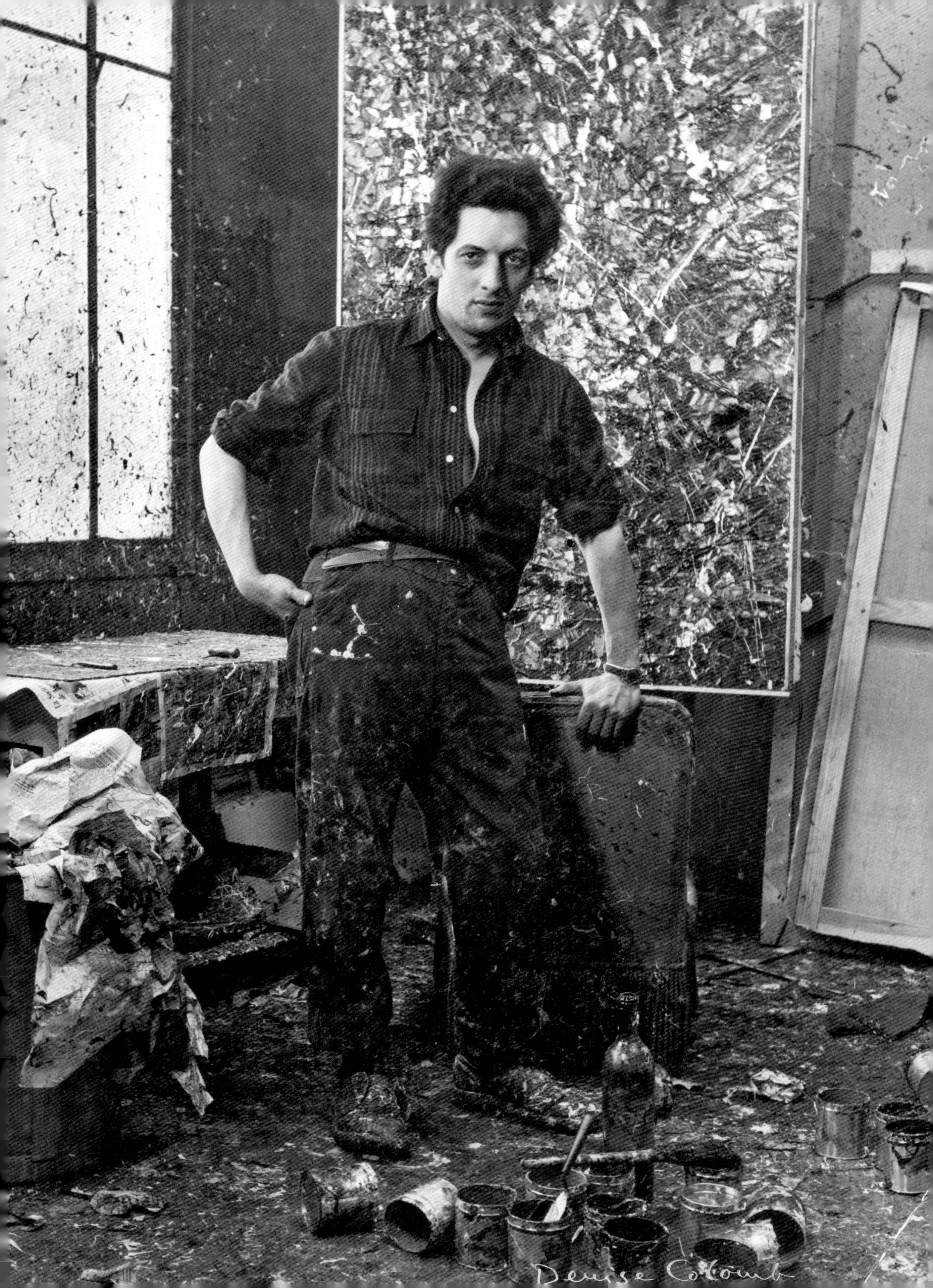
Denise Colomb

Started in the 1880s by Paul Foinet, who used to sell his artisan-made products through door-to-door visits to artists' studios, the business evolved with the opening of a store in 1902 under the name Lefebvre-Foinet, signaling Foinet's association with his son-in-law Lucien Lefebvre (Bond Rafferty 2009).[22] The store, located at the apex of the acute angle formed by the junction of Bréa and Vavin Streets, consequently had two addresses, 2 rue Bréa and 19 rue Vavin, both indicated on the colorman stamp. Lucien was succeeded by his son Maurice, Maurice by his daughter Josette. The store closed in the mid-1990s, but Josette's son continued making paint for a few artists from a location in Brittany for about ten years.

"Foinet's colour is unique," Riopelle said, referring to Maurice Lefebvre-Foinet, who was in charge of the store at the time the artist was living in Paris. "It expresses all the great skill, the experience, the individual presence of one man" (Montreal Museum of Fine Arts 1991, 196). It has been reported that on occasion Riopelle bought all of the paint that Lefebvre-Foinet had on hand and that he sometimes remained idle until specific pigments were once again available (Spalding 2002, 12). Years later, Riopelle expressed disappointment at the quality of paints found on the market, on one occasion saying, "How sad! Before, there was a very rich array of colors" (Viau 1991), and on another, "Who knows how to make colours today?" (Érouart 1995, 22).

Oil Painting Supports

The American art critic Thomas B. Hess made interesting observations in 1979 about Riopelle's oil painting supports:

> To begin with first things first, consider Riopelle's canvases, the supports that define the limits and proportions of the image. Unlike such artists as Pollock, de Kooning, and Kline, who worked on unstretched canvases tacked to a painting-wall or lying on the floor, Riopelle prefers fine linen pulled taut on stretcher frames. For the Americans, a picture's final boundaries are determined towards the end of the creative procedure, when it is almost finished.... In other words, the Americans' pictorial dynamics move centrifugally outwards, from the mid-part of the painting towards edges that are hazy, ill-defined, and only nailed down in a dialectic veering between quavering, intellectual uncertainty and brusque, intuitive decisions.... Riopelle, on the other hand, orders his canvases stretched in certain sizes and proportions that have been an accepted part of French art for some 200 years. When he makes his first mark, in the mid-part of a white surface, he knows where the limits are. There is the *figure* canvas, usually approaching a square. The *paysage* format, usually more elongated, with one axis visibly dominating the other. And there is the *marine*, pulled out, traditionally to accommodate long horizons and sweeps of beach. There is a wide variation within each classification, but—and this should be borne in mind—each is known, familiar, pre-existent. When Riopelle paints, his hand and wrist move within perimeters where, literally, Matisse, Monet, Courbet and his other heroes have moved. Their limits are part of his hypothesis. (Montreal Museum of Fine Arts 1991, 69)

Hess refers to the French system of standard canvas sizes, each identified by a number and available in three formats, *figure*, *paysage*, and *marine*. There is no doubt that Riopelle used such canvases, as many of his paintings have these standard dimensions. The fact

FIGURE 19
Jean Paul Riopelle, *Abstraction*, 1958, oil on canvas, 59.7 × 75.2 cm. Musée national des beaux-arts du Québec, inv. 1995.285. Gift of Jean-Paul and Gilberte Roy. Catalogue raisonné no. 1958.100H.1958.

that Riopelle bought his canvases already mounted on stretchers in the case of standard formats was confirmed by Lefebvre-Foinet (Établissements Lefebvre-Foinet 2002). There are instances in which canvases on stretchers have dimensions that differ from the standard formats but still bear the Lefebvre-Foinet stamp on the back; in these cases, the store may have prepared nonstandard stretched canvases for the artist.[23] He also, however, apparently bought canvas in rolls (Établissements Lefebvre-Foinet 2002).

Monumental works such as *Chevreuse* and *Hommage à Grey Owl*, much larger than the largest standard canvas (*figure* no. 120, 195 × 130 cm), were often painted on one large piece of canvas, as Lefebvre-Foinet supplied pieces of canvas that could be as large as 3 meters wide by 10 meters long (Établissements Lefebvre-Foinet 2002). Other very large formats were constructed of several canvases, each mounted on its own stretcher. An example is *Point de rencontre* (see fig. 6), which measures 428 cm high by 548 cm wide and consists of three vertical canvases of 286 cm by 183 cm placed above two horizontal canvases measuring 142 cm by 274 cm. While the goal of juxtaposing several canvases was often to obtain a larger surface, in other instances it was done to create unusual structures. In *La Mi-été chez Georges*, dated 1975, for example, four small square canvases are located asymmetrically beside a larger vertical canvas, one on the left side and three on the right. Riopelle also painted on oval canvases (fig. 19), on round canvases, and, on occasion, on canvases with an unusual shape, such as that of *Fantaisie*, dated 1961 (Riopelle and Riopelle 2009, 114–15).

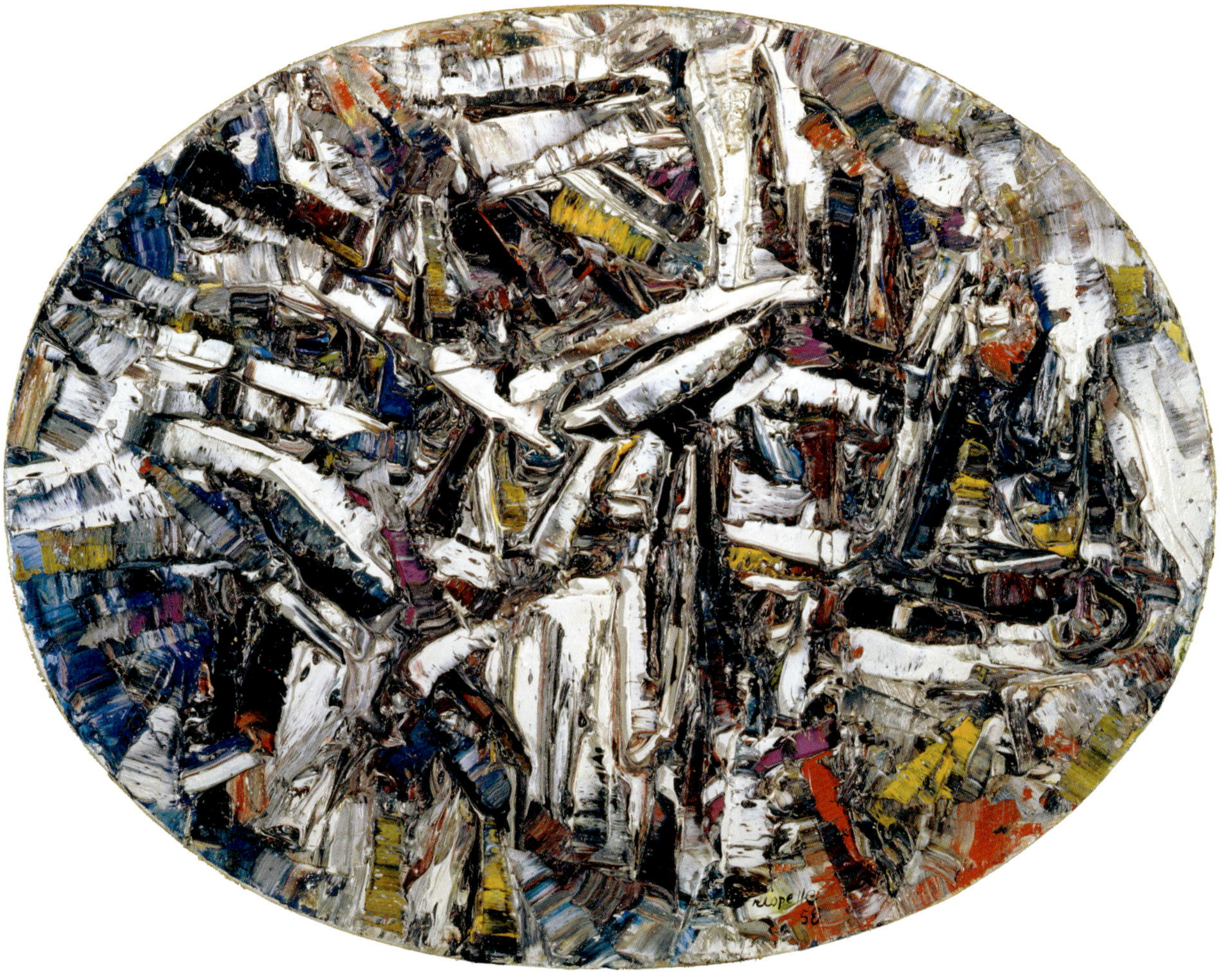

Analysis of samples of canvas by Fourier transform infrared (FTIR) spectroscopy and polarized light microscopy (PLM) showed that Riopelle painted most often on a sized, plain-weave linen canvas. Most canvases examined were relatively coarse (about 12–15 by 10–16 threads per centimeter), although some small-format paintings were painted on a finer canvas (about 20–22 by 20–22 threads per centimeter).

Oil Painting Grounds

Riopelle almost always painted on a white, commercially prepared ground. Among the paintings examined, only one early painting had a ground that was not white but rather brownish beige. There is no evidence that Riopelle ever added a priming or local color over the commercial ground prior to painting.

X-ray diffraction and FTIR spectroscopy showed that most grounds were based on lead white, sometimes consisting of mixtures with zinc white and/or fillers such as calcium carbonate or barium sulfate. A smaller number of paintings had grounds that consisted of a mixture of zinc white and calcium carbonate.

Certain paintings had a lead white ground made up of two varieties of lead white, one known as "synthetic plumbonacrite" (pentalead tricarbonate dihydroxide oxide) (Corbeil and Sirois 2007) and the other the usual hydrocerussite type (trilead dicarbonate dihydroxide). This type of ground—the plumbonacrite-hydrocerussite combination—was used by Lefebvre-Foinet and has been found in paintings by other artists who bought their supplies there, such as Pierre Soulages (Helou-de la Grandière, Le Hô, and Mirambet 2008) and Paul-Émile Borduas (Challan-Belval 1991).[24]

Analysis by FTIR spectroscopy and gas chromatography-mass spectrometry (GC-MS) identified the binding medium of all the ground samples as oil, in most cases linseed oil (see Appendix II for more details).

Oil Paint

Riopelle stopped using brushes early in his career; instead he applied paint directly from the tube using a palette knife.[25] Although no one was asked into his studio, he reportedly had a witness when he was working in an old barn on Long Island during the summer of 1960. The owner of the barn described Riopelle at work:

> I will never forget this scene. First, he did not paint with a brush but rather with what looked like a putty knife. Second, judging by the hundreds of empty tubes that lay at his feet, he was using a phenomenal quantity of paint. He did not unscrew his tubes. He decapitated them in one move with his knife without ever using the cap. Red, blue, or green: the colors appeared suddenly at the tips of his fingers. Because that is how he was doing it: he held all the tubes (say three or four or as many as his hand could hold) in his fist and then either poured them directly on the canvas or managed to have one color mixing with the next by pressing the tubes in a certain way. (de Billy 1996, 155)

Indeed, photographs taken in the 1950s show Riopelle next to paintings from the *Mosaïques* series with a huge pile of used paint tubes at his feet (see fig. 2),[26] and examination of tubes from his studio in France confirmed that they were indeed "decapitated."

FIGURE 20
Cross section prepared from a sample from *Un Coin de pays*, showing superimposed layers of unmixed paint.

Photo © Canadian Conservation Institute

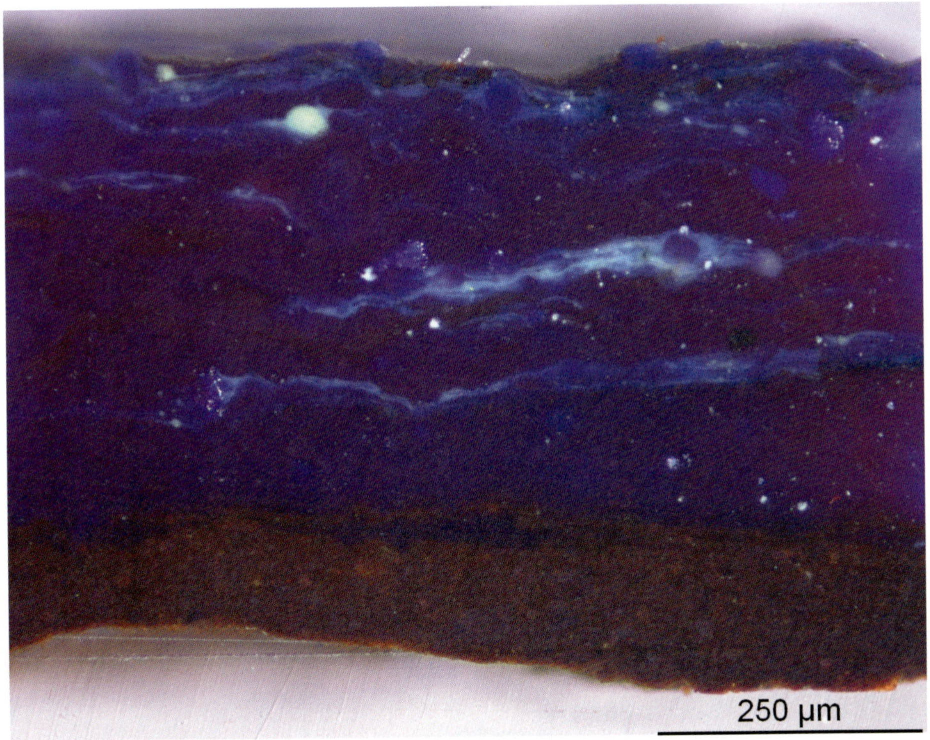

In his paintings from 1949 and the early 1950s, the presence of long filaments of paint has often been seen, incorrectly, as resulting from a dripping technique like Pollock's. According to Riopelle, these filaments came from having a large load of paint on the palette knife, although he admitted that he sometimes used a brush or threw paint to create such effects (Letarte and Feaver 1982).[27] Furthermore, Riopelle was working on an easel with the painting in a vertical position, or, for an oversize painting, by leaning the stretched canvas against the wall; he never worked with his canvases on the floor, therefore making "dripping" virtually impossible.

As a consequence of his working method, paint was not mixed before being applied; instead, it was applied directly from the tube and then worked onto the canvas. While paint could become intermixed as strokes of paint were applied with the palette knife, in many cases little mixing actually took place, as cross sections prepared from samples of the oil paintings show (fig. 20). Rich, vibrant, saturated colors are found in Riopelle's oil paintings. Analysis of paint samples reveals that most contained only one pigment, sometimes with fillers, and many samples have compositions identical to those of the paint tubes, indicating that Riopelle used unmixed paints.

Pigments

Riopelle favored specific pigments for the majority of his colors, although analysis reveals the sporadic use of other pigments. The inventory of paint tubes from his studio in France supports this observation. Riopelle admitted that he never began a painting "without having ready to hand all the colours" he would need. He added, "I would even go so far as to say—obviously I don't use a palette, but the idea of a palette or a selection of colours that is not *mine* makes me uncomfortable, because when I work, I can't waste my time searching for them. It has to work right away" (Érouart 1995, 79–80). Indeed, when Riopelle started painting, he did it in one session, without interruption, unaware of time passing.

The pigments in Riopelle's works were identified with a combination of X-ray diffraction, FTIR spectroscopy, and in some cases PLM. X-ray diffraction was used to identify most of the inorganic pigments and fillers; FTIR spectroscopy was used for most of the organic pigments and some of the inorganic pigments and fillers. In some cases, microsolvent extractions with acetone or chloroform were used to separate the organic pigment from the paint matrix.

Cadmium-based pigments (cadmium sulfide and selenosulfide pigments) were most frequently used for yellow, orange, and bright red shades. Chrome yellow was also used often, as well as yellow and red organic pigments such as Hansa yellow 10G (PY 3),[28] toluidine red (PR 3), and chlorinated para red (PR 4). Riopelle made frequent use of deep burgundy and violet, consisting of alizarin red (PR 83) and cobalt violet, respectively. Analysis shows that the cobalt violet pigment was more specifically of the magnesium cobalt arsenate variety.[29]

Riopelle used black and white in almost every painting. Bone black was used almost exclusively for black, while lead white was the most frequently used white pigment. Other white pigments, such as zinc white and titanium white, were used mostly in works dating from before 1952 and from the late 1970s. Both zinc white and titanium white were used in paintings from the black-and-white series *Icebergs*, such as *Soleil de minuit (Quatuor en blanc)* (see fig. 13) and *Pangnirtung* (see fig. 9), both dated 1977. Of the ninety-two paint tubes from Riopelle's studio in France, there were fifteen bone black tubes (labeled "noir d'ivoire") and twenty-six lead white tubes (labeled "blanc d'argent") bearing a Lefebvre-Foinet label, as well as seven tubes resembling Lefebvre-Foinet tubes but with no label, also containing black or white paint.

Some samples that appeared black were composed of Prussian blue or ultramarine blue, which, undiluted, are a deep blue, almost black (fig. 21). Others consisted of mixtures of bone black and Prussian blue, ultramarine blue, or viridian.

FIGURE 21
Cross section prepared from a sample from *Envol*: the thick bottom layer appears black but consists of Prussian blue; the thin white layer underneath corresponds to the ground.

Photo © Canadian Conservation Institute.

Several blue pigments were identified in Riopelle's oil paintings. Cobalt blue and ultramarine blue were found most frequently. Prussian blue, cerulean blue, and phthalocyanine blue (PB 15) were also sometimes used. Green in his paintings was often obtained by using viridian but was also achieved by mixing green and yellow pigments or by mixing green, yellow, and blue pigments. It is not possible to determine with certainty if the mixtures are those of the colorman or if they are Riopelle's, but given Riopelle's technique, with very little mixing taking place, the first hypothesis seems more likely. Phthalocyanine green (PG 7) was also identified in some paintings.

Medium

Riopelle's oil medium was analyzed by FTIR spectroscopy and GC-MS. Detailed results of the analysis are provided in Appendix II. As his method of applying paint directly from the tube would lead one to expect, there was no indication from the analysis that Riopelle mixed his oil paint with other paint media. A distribution of the palmitate to stearate (P/S) ratios measured in the oil paint samples is given in Appendix II. Many of Riopelle's paints have a P/S ratio below 2.0, consistent with the use of linseed oil, while a number of samples show P/S ratios well above 3.0, indicating the use of poppyseed oil (Mills and White 1994, 33). The many P/S ratios between 2.0 and 3.0 are difficult to interpret. Analysis of the full fatty acid profile of wet paint from a paint tube from his studio shows that Riopelle did use paints based on safflower oil, which has a P/S ratio in this intermediate range. However, as Appendix II explains, intermediate P/S ratios may also be obtained from other oils or mixtures.

Varnish

Riopelle did not varnish his paintings, but unfortunately some are now varnished. Samples of varnish were taken when it was observed on paintings. The analysis of the varnishes showed that Riopelle's paintings have been treated both with acrylics, which would tend to have a more matte effect, and with ketone resins and natural resins, which produce a glossier surface. Neither treatment is acceptable. Both alter the contrast between matte and glossy paint that the artist sought, as discussed in chapter 5.

4 Exploration of Mixed Media

In their 1968 interview, Seguin commented that during a time when modern painting had seen a number of technical developments, such as the introduction of acrylic paint, Riopelle remained faithful to oils. "My way of looking at a painting and judging it," Riopelle replied, "corresponds to the way paintings created in oil on an easel have always been seen and judged; the idea of painting on canvas pleases me more than the idea of experimenting with another material, whether harder or softer; that interests me less" (Érouart 1995, 80).

In the 1970s, however—just a few years after making this comment—Riopelle began to use acrylic paint. He also integrated gold leaf in his works as early as 1976 (Bernier 1997, 135; Montreal Museum of Fine Arts 2002, 114). But the true revolution in his art occurred in the 1980s, when he started using nontraditional materials such as aerosol paint and paint for crafts and fabrics. Then in 1989 he discovered paints with integrated glitter, both acrylics and aerosols (Vachon 2000, 95), and also made use of gold, silver, and copper leaf in works produced that year (Riopelle 2010c). He had finally reached the stage of experimenting with new materials, something he had had no appetite for when he was painting in oil.

But the use of new or nontraditional materials did not mean that Riopelle was no longer interested in the quality of his paint. He took pride, for example, in showing some of his materials to Huguette Vachon, who became his partner in 1987: "These felts [*sic*] are from New York. They're permanent. They can stand full light" (Vachon 2000, 95).

Access to good-quality materials in sufficient quantities was often a problem for Riopelle, and it became even more of an issue when he installed his studio on Île-aux-Oies in the fall of 1990. The location was perfect for his work, but finding proper artists' materials was problematic. To Érouart, who asked Riopelle if working on an island was different, Riopelle replied:

> An island is like anywhere else, and it's always different. I have to get off to a good start, know where to attack, before I can really begin to paint. After that, whether I'm on an island or on *terra firma*, time no longer exists.... When I paint, it's no longer the sun that marks out the day, but technical things. The colours I've ordered that haven't arrived, the paper I crave but that you can only find in Paris.... Far too often, I'm disappointed that I can't get the materials I once could. (Érouart 1995, 22)

Riopelle reportedly bought all the doors he could find in stores in the neighborhood, as well as car aerosol paint (Bertrand 1991). The doors were used to paint a series of double-sided panels, such as *Matinée au Cap Tourmente (Les Faisans dans la volière)/L'Oie hélico* and *Hommage à Duchamp (Hommage à Maurice Richard)*, two paintings included in the CCI study.[30]

Good-quality canvas may also have been hard to get. Huguette Vachon wrote in her diary: "Jean-Paul needs canvas. Last week, a little bit for us and to remind us of Joan [Mitchell], I bought a roll of canvas. Of course, mediocre quality. 15 meters. Jean-Paul finished it last night" (Bernier 1997, 66).

Supports and Grounds

While all oil paintings produced by Riopelle were executed on canvas, he explored a variety of supports for his mixed-media works. The ones included in CCI's study of the materials and techniques (table 3) were an acrylic on lithograph (adhered to canvas) (*F-de*, 1975), two acrylic and ink on serigraph (adhered to canvas) (*Innocente* and *Envol (Les deux soleils)*, both dated 1985), two mixed-media on wood panel painted recto and verso (*Matinée au Cap Tourmente (Les Faisans dans la volière)/L'Oie hélico* and *Hommage à Duchamp (Hommage à Maurice Richard)*, both dated 1990), and the large-scale mixed-media triptych on canvas, *L'Hommage à Rosa Luxemburg*.

According to Harry Bellet, the paintings in the series painted on doors in lieu of wood panels generally had a matte aspect because the panels were not primed (Bellet 1991, 41). Analysis showed that this was not always true, as a white acrylic ground was found on *Matinée au Cap Tourmente*. No ground layer was observed in the case of *Hommage à Duchamp*.

The canvas of *L'Hommage à Rosa Luxemburg* could not be safely sampled and therefore was not analyzed. It was observed to be commercially prepared with a white ground. Was it one of the "mediocre"-quality canvases Huguette Vachon mentioned in her diary? Each canvas of the triptych is fifteen meters long, like the roll she described.

TABLE 3 Mixed media included in the study

Title	Year	Technique	Dimensions (cm)	Collection*	Accession number
F-de	1972	acrylic on lithograph adhered on canvas	86 × 46	MNBAQ	90.291
Innocente	1985	acrylic and ink on serigraph adhered on canvas	about 75 × 108	MMFA	97.2002
Envol (Les deux soleils)	1985	acrylic and ink on serigraph adhered on canvas	about 75 × 108	MMFA	96.2002
Matinée au Cap Tourmente (Les Faisans dans la volière)/L'Oie hélico	1990	mixed media on wood panel painted recto verso	124.5 × 246.5	MMFA	2001.165.1-2
Hommage à Duchamp (Hommage à Maurice Richard)	1990	mixed media on wood panel painted recto verso	203.2 × 91.4	MACM	D 94 35 P 1
L'Hommage à Rosa Luxemburg	1992	mixed media on canvas	155 × 4039 (3 canvases; from left to right: 155 × 1424; 155 × 1368; 155 × 1247)	MNBAQ	96.96

* MACM: Musée d'art contemporain de Montréal (Canada); MMFA: Montreal Museum of Fine Arts (Canada); MNBAQ: Musée national des beaux-arts du Québec (Canada).

Paint

Method

Aerosol paint was usually used for the background of the paintings. Objects such as nails, nets, horseshoes, feathers, leaves, and birds were often placed on the support and sprayed; once the objects were removed, negative impressions of their shapes remained (fig. 22; see also figs. 8, 10, and 14). The effect of this technique is somewhat reminiscent of prehistoric rock paintings.[31] As Bellet has noted, "Magdalenian artists signed the animals they drew by pressing a paint-covered hand alongside on the wall, or by blowing paint onto their hand through a tube. These latter negative images are very similar to those that Riopelle projects onto paper using spray paint and stencils. The shape may be cut out of cardboard or taken from an actual object" (Bellet 1991, 39). Lines, dots, and various designs were added using acrylic paint or paint for crafts and fabrics directly from a tube or squeezable bottle (figs. 23, 24). Entire designs were sometimes stamped on the support. For this technique, an object with a hard surface—such as a section of a tree trunk—was painted, then placed against the support, left to dry a little, and finally pulled off. The resulting paint has an unusual texture related to the method of application. Riopelle often used nacreous or metallic paints. As in his oil paintings, he continued to create works in which color, volume, and range of gloss played an important role.

Pigments and Medium

Samples of aerosol, nacreous, and metallic paints as well as more conventional paints were analyzed by the techniques described in Appendix I. The analysis of the aerosol paints was hindered by the fact that they were applied very thinly and that several of the samples were contaminated by varnish. However, in a number of cases the medium and pigments were identified. In *L'Oie hélico*, yellow and blue aerosol paints were found to have an oil-modified iso-phthalate alkyd resin medium. Aerosol paints are typically nitrocellulose, acrylic, or alkyd paints (Marchbank 1974; Béguin 1981, 24). Owing to the limited number of samples of aerosol paints analyzed and the difficulties mentioned above, it was not possible to determine whether Riopelle used a range of aerosol paint media.

In *L'Hommage à Rosa Luxemburg* and *Hommage à Duchamp (Hommage à Maurice Richard)*, red and orange fluorescent aerosol paints contained a high proportion of a melamine-formaldehyde-sulfonamide resin, which acts as a carrier for the fluorescent pigments. Most organic fluorescent pigments are based on either xanthene or amino-naphthalimide dyes immobilized in a carrier composed of triazine-formaldehyde-sulfonamide resin (Voedisch 1973). Other pigments found in the aerosol paints were Hansa yellow (probably PY 74) and phthalocyanine blue (PB 15). Gold and silver aerosol paints were colored by brass powder and aluminum powder respectively.

As well as the aerosol paints described above, other gold and silver metallic paint consisted of small, thin squares of aluminum suspended in an acrylic-acrylonitrile-styrene copolymer medium. In the silver paint, the aluminum flakes were mixed into a colorless medium, while gold paints were produced by mixing the aluminum flakes into a yellow-tinted medium.

The same acrylic (acrylic-acrylonitrile-styrene copolymer) was found in most of the nonaerosol paint samples. The pigments identified in these paints included organic pigments such as phthalocyanine blue (PB 15), benzidine yellow (PY 14), dioxazine violet (PV 23), and a naphthol red. White paint was pigmented with titanium white (rutile) and black paint with a finely divided carbon pigment. A type of red nacreous paint was also found, consisting of an interference pigment suspended in the acrylic medium. Interference pigments are composed of platelets of natural mica (usually

FIGURE 22A
Detail of Jean Paul Riopelle, *L'Hommage à Rosa Luxemburg*, 1992. Part 2, panel 5. Musée national des beaux-arts du Québec, accession no. 96.96. Gift of the artist.

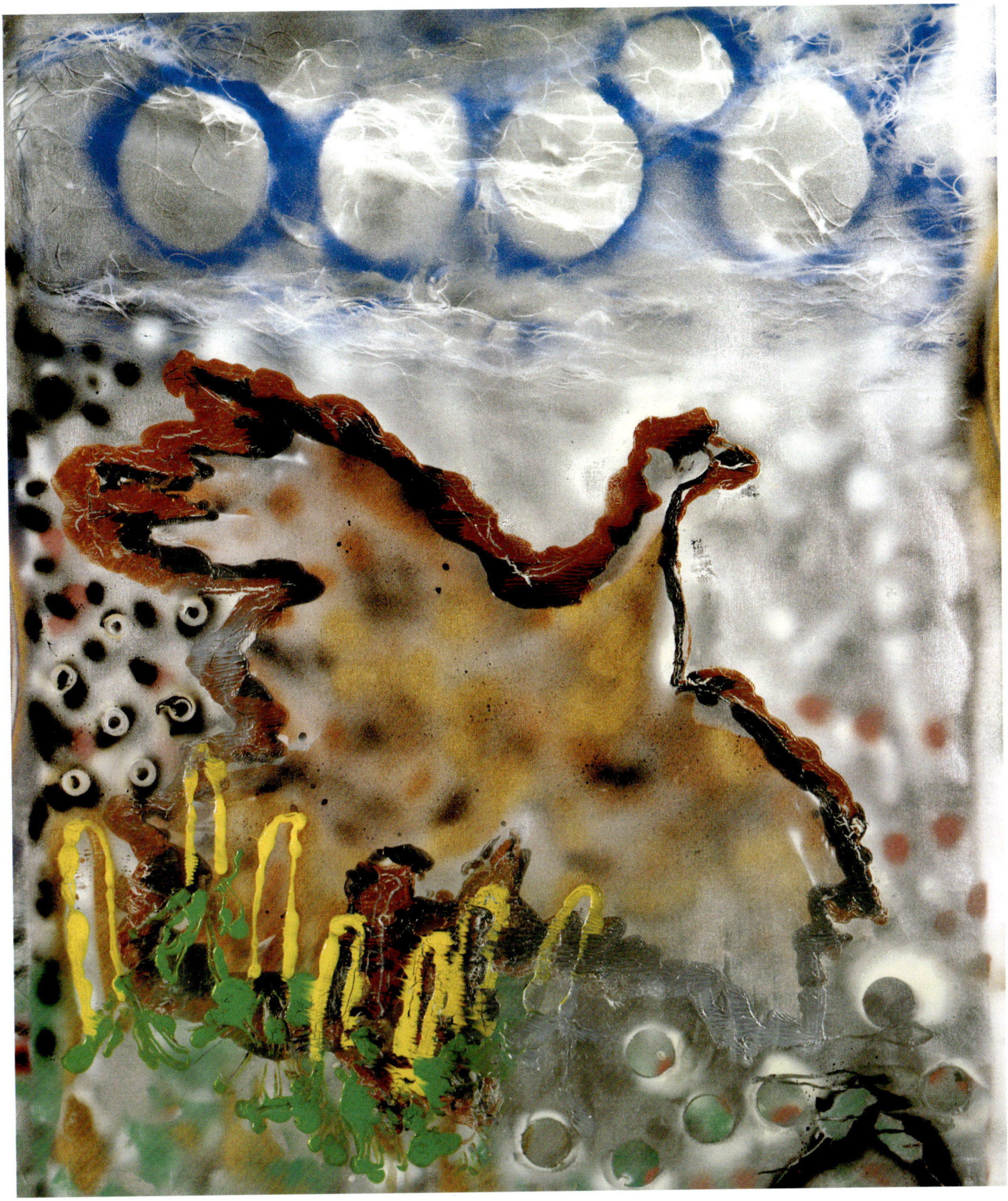

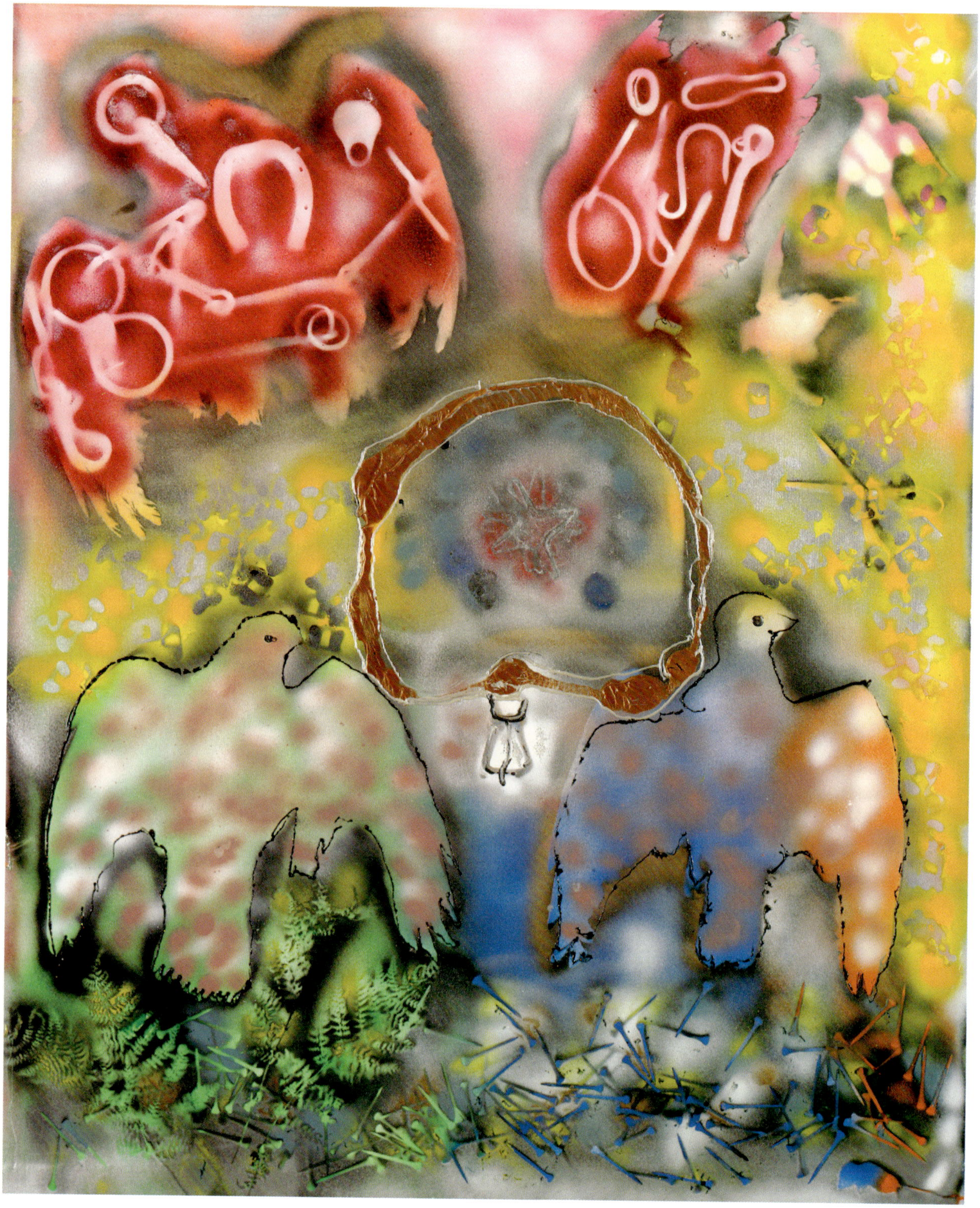

FIGURE 22B–C
Details of Jean Paul Riopelle, *L'Hommage à Rosa Luxemburg*, 1992. (B) Part 3, panel 8. (C) Part 3, panel 1. (Part 3 reads from right to left.) Musée national des beaux-arts du Québec, accession no. 96.96. Gift of the artist.

Photographer: Patrick Altman

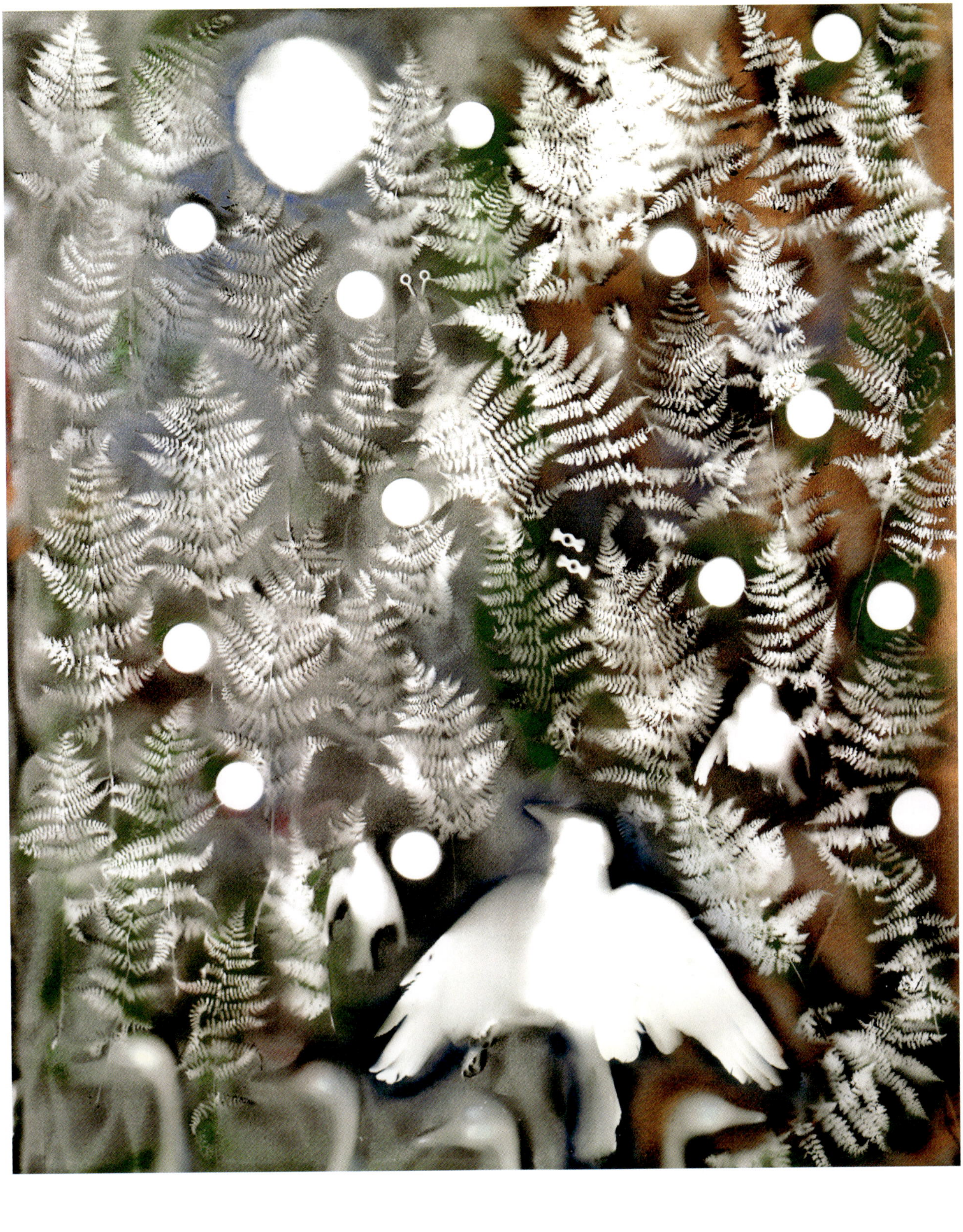

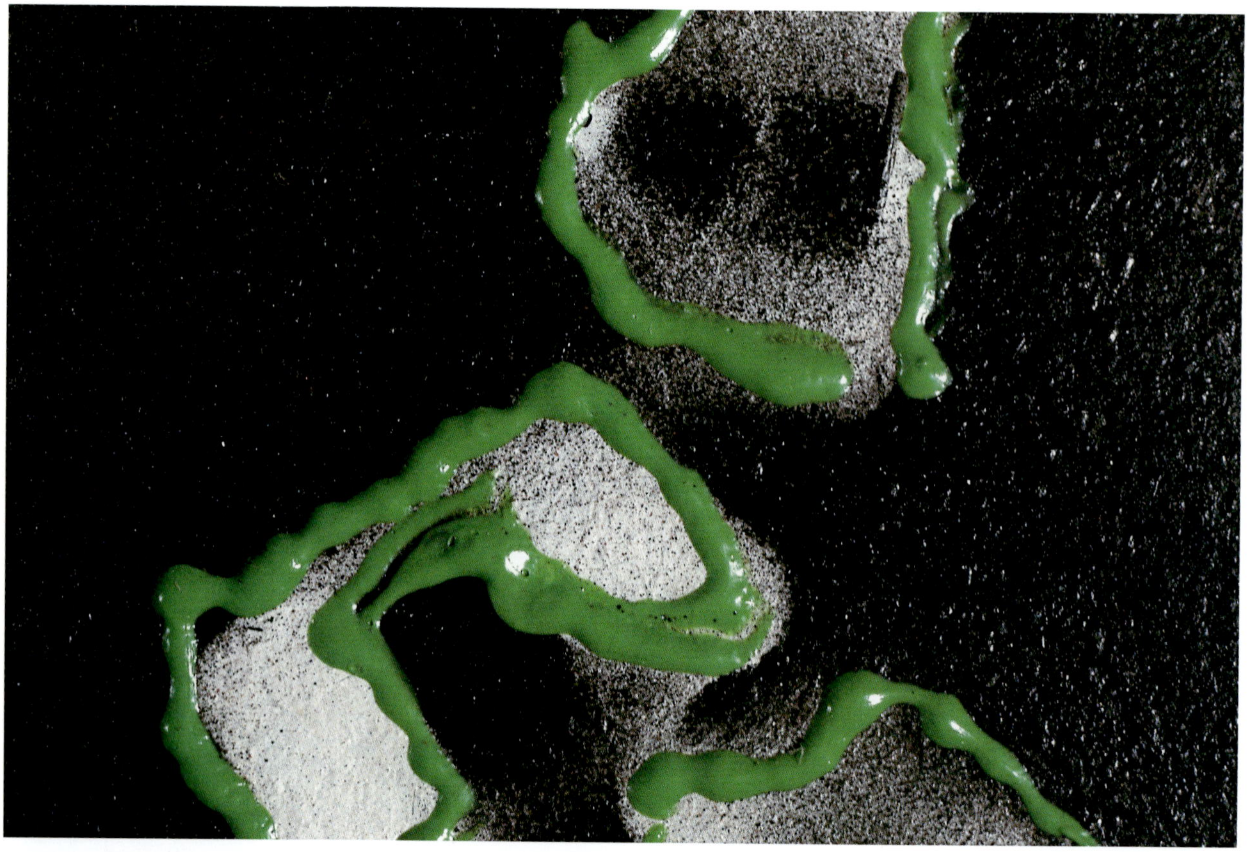

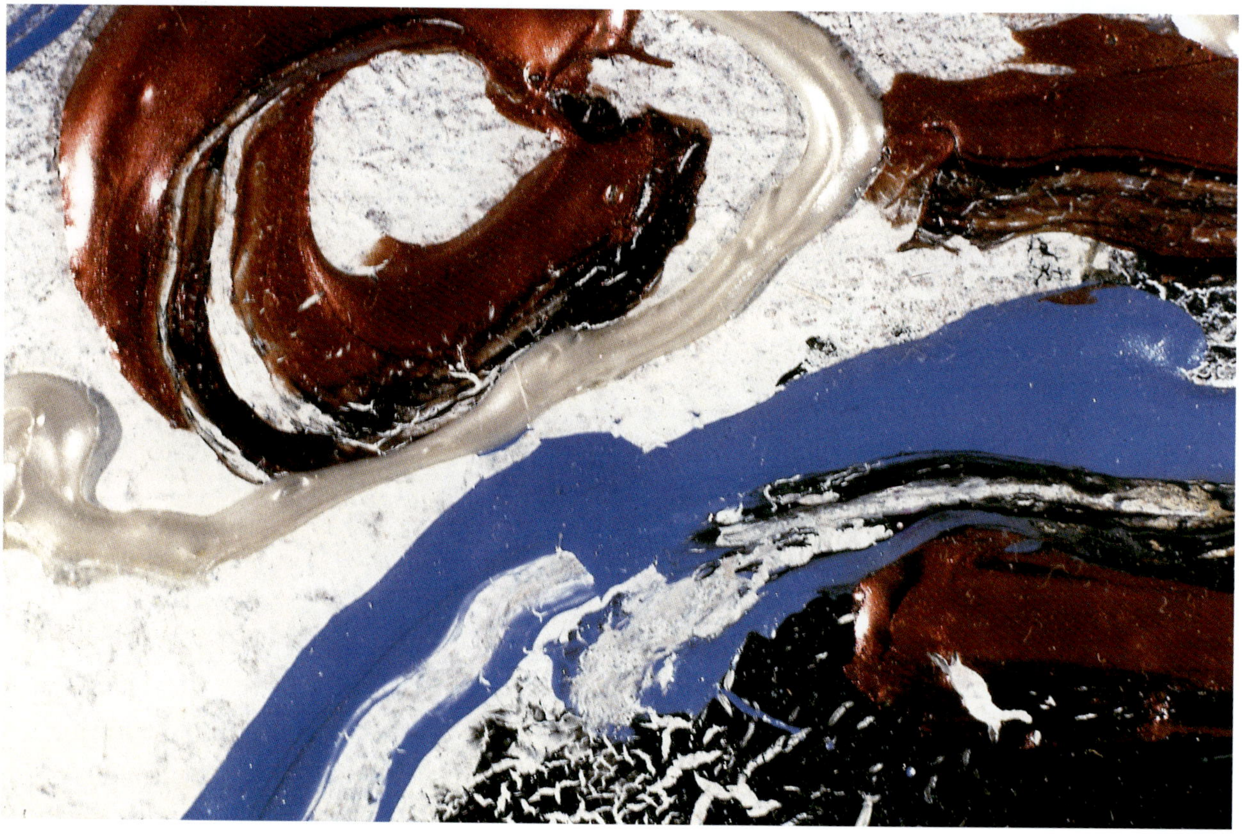

FIGURE 23
Detail of *L'Oie hélico* showing green paint used to create contours against a black aerosol paint background.

Photo © Canadian Conservation Institute.

FIGURE 24
Detail of *L'Oie hélico* showing red nacreous paint used to paint the eye and beak of the goose.

Photo © Canadian Conservation Institute.

FIGURE 25A–B
Cross section prepared from a sample of red nacreous paint from *L'Oie hélico*, showing platelets of interference pigment suspended in an acrylic copolymer: (A) photomicrograph; (B) backscattered electron micrograph.

Photos © Canadian Conservation Institute.

FIGURE 26
Dispersion of a sample of bronze-colored nacreous paint from *L'Hommage à Rosa Luxemburg*, showing interference pigment with green-red flop.

Photo © Canadian Conservation Institute.

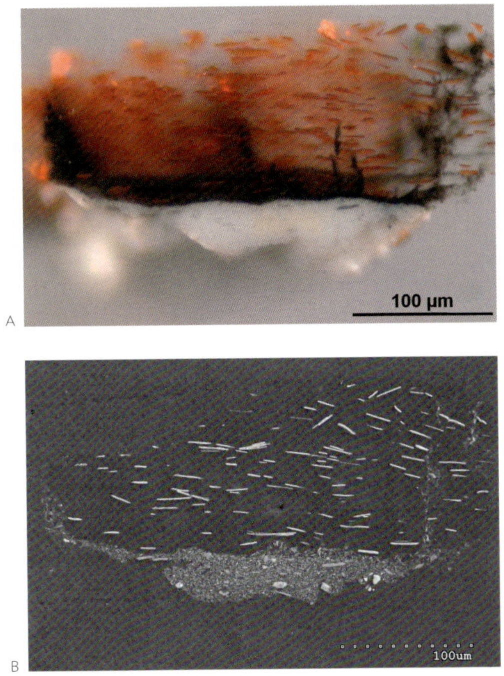

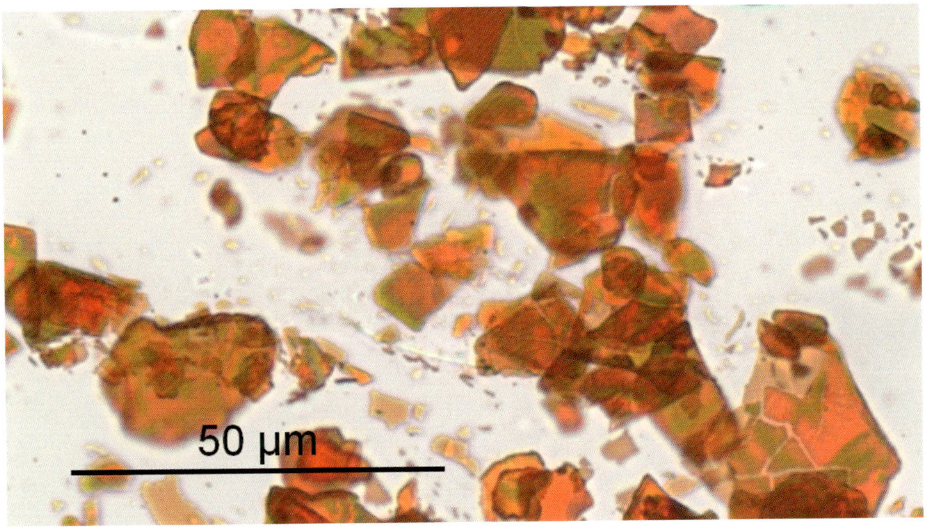

muscovite) coated with thin films of transparent metal oxides (Franz et al. 1993, 214), iron oxide in this case. A cross section prepared from a sample of this paint is shown in figure 25. A second type of nacreous paint with a brown color (which produced almost a bronze finish) was found to have a similar composition. In this case, the dispersion observed by PLM showed flakes that have two colors, orange and olive green (fig. 26). In iron oxide–mica pigments, green-red color flop (that is, a change in color depending on the viewing angle) is produced by depositing an iron oxide layer with a specific thickness that produces green interference colors (Franz et al. 1993, 217).

Acrylics found in the reworked lithograph and silkscreens were based on an ethyl acrylate–methyl methacrylate copolymer. The pigments identified in these paints included ultramarine blue, phthalocyanine blue (PB 15), phthalocyanine green (PG 7), cadmium yellow, and Hansa yellow G (PY 1). White paint was pigmented with titanium white (rutile) and black paint with a finely divided carbon pigment.

5 The Painted Oeuvre: Observations on Condition and Conservation

Riopelle's unorthodox working methods, including his use of the traditional oil medium in nontraditional and unconventional ways and, later, his adoption of mixed media and commercial materials, pose challenges for the treatment and preservation of his works. As part of the research project on Riopelle's painting technique, a number of the artist's paintings were examined, conservation files were consulted, and conservators who had previously worked on paintings by Riopelle provided information on what they had observed and how they had treated the works. The first two sections of this chapter focus on conservation problems frequently encountered with Riopelle's paintings—problems rooted in the artist's materials and technique. The third section addresses certain physical characteristics of the works that must be taken into account in order to preserve the aesthetic qualities of Riopelle's oil paintings, in particular when developing a conservation treatment.

Common Conservation Issues Observed in Oil Paintings

Several conservation problems frequently observed in Riopelle's oil paintings are the lifting and flaking of paint, the presence of fragile, cracked, or soft paints, and the appearance of exudates.

Flaking/Lifting Paints and Soap Formation

FTIR spectroscopy showed that most of the grounds and many of the paint samples in Riopelle's oil paintings contained lead or zinc soaps. Details regarding the characterization of the soaps in the paint layers by FTIR spectroscopy are given in Appendix II. Metal soaps, also referred to as metal fatty acid salts, are frequently found in drying oil paint (Higgitt, Spring, and Saunders 2003; Robinet and Corbeil 2003). Sometimes metal soaps such as aluminum or zinc stearate were added to paint by the manufacturer to improve its handling and dispersing properties (Mayer 1991, 138–39; Learner 2004, 21–22); in other cases the metal soaps found in paint are the result of a reaction between metal ions in the pigment and fatty acids in the oil medium. Lead white and zinc white readily form metal soaps in oil paints. While the presence of metal soaps in itself is not necessarily an issue, high concentrations of soaps have been linked to problems with the paint film.[32] A high concentration of soaps would be expected to affect the cohesion and mechanical properties of the ground and/or paint layers and could lead to a lack of adhesion between layers. In some cases, paint layers with high concentrations of soaps have been described by conservators as having a waxy texture (Moffatt and Miller

FIGURE 27
Detail of *Point de rencontre* showing flaking occurring between the burgundy red paint and the ground.

Photo © Canadian Conservation Institute.

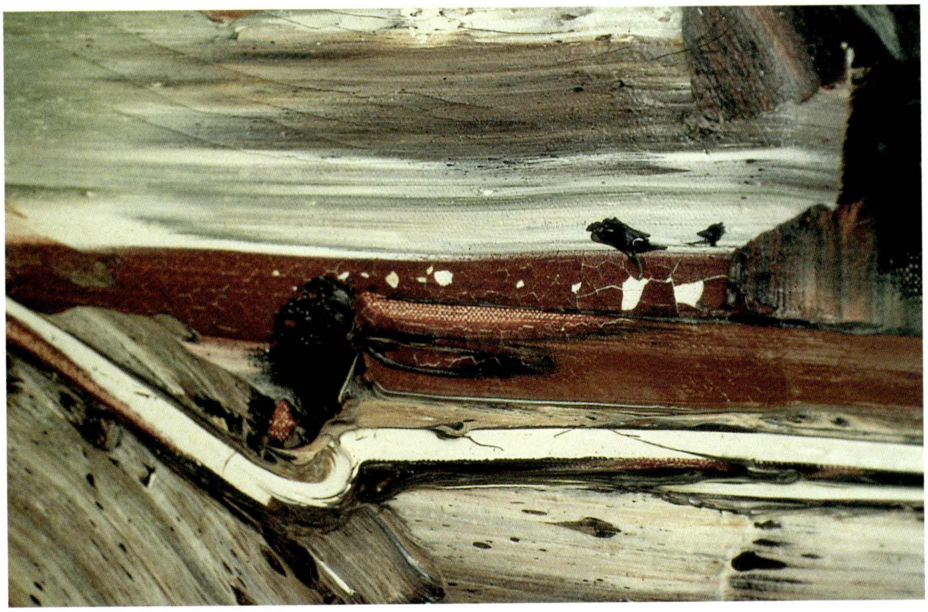

1994; O'Malley and Moffatt 2001).

Lifting and flaking of paint in Riopelle's works occurs often, most commonly at the ground/paint interface (fig. 27). Because the particular type of lead white used in prepared canvases supplied by Lefebvre-Foinet could be a factor in the lack of adhesion between the ground and the paint layers, the degree of soap formation in the lead white grounds was assessed by examining the FTIR spectra. As mentioned in chapter 3, some of these grounds contain the plumbonacrite form of lead white (pentalead tricarbonate dihydroxide oxide) in addition to the usual hydrocerussite form (trilead dicarbonate dihydroxide). Since the plumbonacrite form of lead white most likely has a different basicity from the hydrocerussite form, it may have an influence on the drying of the oil and the formation of soaps.

Small amounts of lead soaps, identified by a shoulder or a weak band at approximately 1515–1520 cm^{-1} (the position of the most intense asymmetric COO^- stretch for lead soaps; Robinet and Corbeil 2003), were found in most of the lead white grounds. A few of the ground samples showed a more intense band at 1515 cm^{-1}, indicating the presence of a significant quantity of lead soaps. The ground samples containing plumbonacrite did not contain a higher concentration of lead soaps than the other lead white grounds. Lead soaps do not necessarily occur uniformly throughout a paint layer, however. For instance, the analysis of ground samples from several locations on the painting *La Mi-été chez Georges*, a painting with a lead white ground that did not contain plumbonacrite, shows that the concentration of lead soaps varies with location. Figure 28 shows the FTIR spectra of the lead white ground from this multicanvas painting in two locations on the same panel. The spectrum in the upper trace of the figure shows almost no lead soaps, while the spectrum in the lower trace shows a high concentration of soaps.[33] Study of cross sections in different areas of the paintings would be needed to fully understand the distribution and amount of lead soaps in the two types of lead white ground layers.

Similar lifting of paint at the ground/paint interface was observed in paintings by Pierre Soulages, who also purchased his materials from Lefebvre-Foinet. As part of a study of his paintings from the late 1950s, Soulages was interviewed; he mentioned having used "non-absorbent pre-primed canvases from Lefebvre-Foinet. These grounds

FIGURE 28
FTIR spectra of ground samples from different locations on the painting *La Mi-été chez Georges*. The upper trace shows almost no lead soaps, while the lower trace shows a high concentration of lead soaps.

Photo © Canadian Conservation Institute.

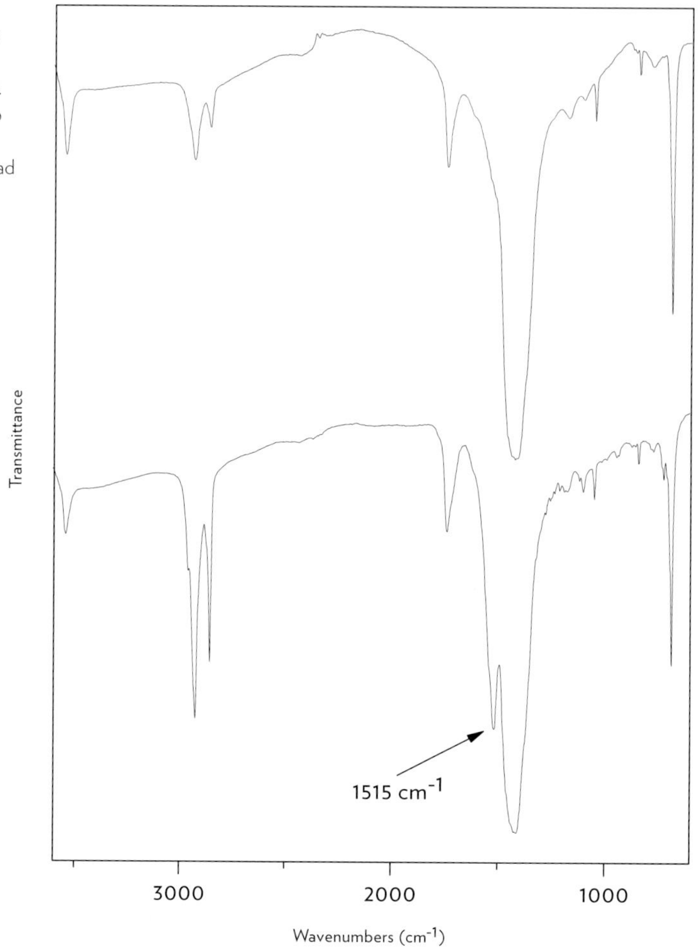

made of lead white were chosen because they were more 'greasy' than others, and they prevented the paint medium from being absorbed by the ground" (Helou-de la Grandière, Le Hô, and Mirambet 2008, 157). Analysis of samples from these paintings showed that the ground from damaged paintings consisted of two types of particles, described as "acicular" and "round," while the ground from an undamaged painting consisted only of rounded particles (Helou-de la Grandière, Le Hô, and Mirambet 2008, 159). Ground samples from some of the paintings containing the two types of particles were further investigated by X-ray diffraction and identified as a mixture of the plumbonacrite and hydrocerussite forms of lead white (Corbeil and Helwig 2008). Along with the differences in chemical properties between the plumbonacrite and hydrocerussite forms of lead white, the particular morphology of the plumbonacrite form may affect the drying of the oil in these ground layers: typically in the form of platelets, the plumbonacrite form may not mix with oil in the same way that the usual lead white, generally composed of small, rounded particles, does.

Like Riopelle and Soulages, Paul-Émile Borduas purchased his material from Lefebvre-Foinet when he lived in Paris in the 1950s. His paintings from this period, like theirs, show areas of lifting paint (Challan-Belval 1991). In the case of Borduas's paintings, however, high concentrations of lead soaps were observed in the white paint directly on the ground, and lifting occurred between this white underlayer and the black paint above (Moffatt and Miller 1994), rather than between the ground and the paint as in Riopelle's works.

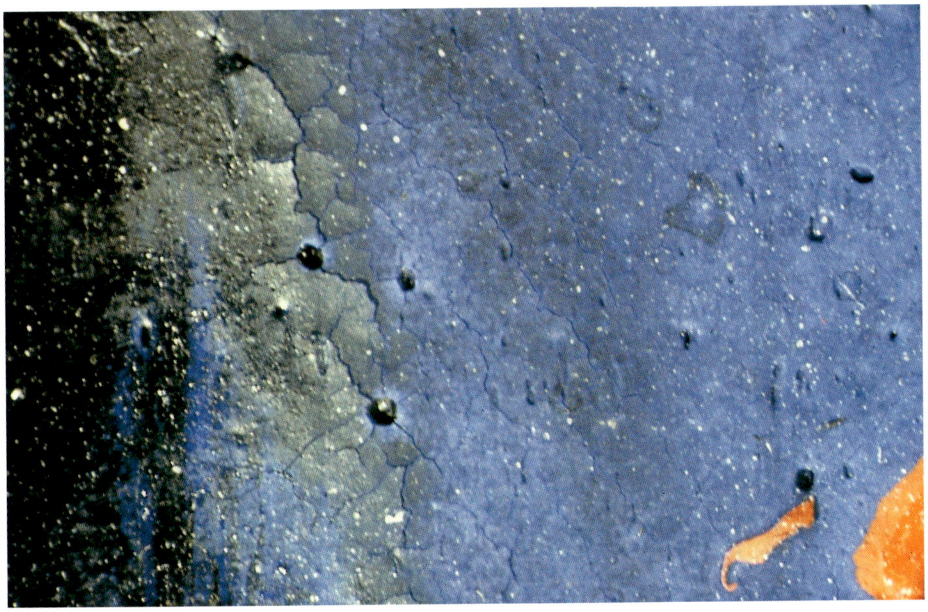

FIGURE 29
Detail of *Point de rencontre* showing powdery, cracked blue paint.

Photo © Canadian Conservation Institute.

Fragile, Cracked, and Soft Paint

Strokes of fragile, powdery blue paint were observed on a number of the oil paintings examined (fig. 29). In other works, the blue paint was particularly shiny or even still soft. Some paintings showed both matte and glossy areas of blue paint in the same brushstroke, as if the paint was not well mixed. Black paint was also sometimes cracked or soft, although less often than blue. The other colors were typically in good condition.

Samples of blue paint from twenty-six oil paintings dating from 1944 to 1973—including blue paint that was well preserved; powdery, matte, or cracked; and shiny or soft—were analyzed. Of the nine samples of problematic blue paints, five were pigmented with cobalt blue and four with another type of blue pigment. Six of the nine problematic blue paints had a palmitate to stearate (P/S) ratio above 2.0, suggesting the presence of a slow drying oil such as safflower or poppyseed. There is some indication, then, that the problematic blue paints are more often pigmented with cobalt blue and contain a slow drying oil. There is no clear correlation, however, as the same pigment/oil combination was found in samples of blue paint that showed no defects. Adding to the difficulty of reaching a clear conclusion is the fact that determining whether a sample is intact or problematic involves a subjective interpretation.

To investigate this further, the azelate to palmitate (A/P) ratio for the various colors of paint was studied by GC-MS. Azelaic acid is the principal diacid that is formed as the oil dries and ages (Mills and White 1994, 39–40; Colombini, Modugno, and Ribechini 2009, 198), and high levels can be indicative of excessive chain scission reactions, which can result in weaker, less cross-linked films (Miller and Moffatt 1994). The detailed results of the A/P ratios for each color of paint are presented in Appendix II. The results show that the blue paints, and to a lesser extent the black paints, have a significant proportion of samples with elevated A/P ratios. This suggests that the problematic paints are related to excessive chain scission reactions, leading to a weaker paint film.

When the blue paint samples were separated by pigment type, the analysis showed that 85 percent of the cobalt blue samples had an elevated A/P ratio. The very high oil absorption index for cobalt blue (Roy 1997, 157) may be a factor in the problematic drying of these paints. Bone black, found in a smaller number of paints with an elevated A/P ratio, is a very slow drying pigment and also has a high oil absorption index (Mayer 1991, 133, 136). In almost all the blue and black paint samples with elevated

FIGURE 30A–B

(A) Jean Paul Riopelle, *Mitchikanabikong*, 1975, oil on canvas, triptych, 195.5 × 391.5 cm (3 canvases, 195.5 × 130.5 each). Musée national d'art moderne (France), inv. AM 1979-254. Catalogue raisonné no. 1975.005H.1975.

(B) Detail showing brown-colored exudates on white paint.

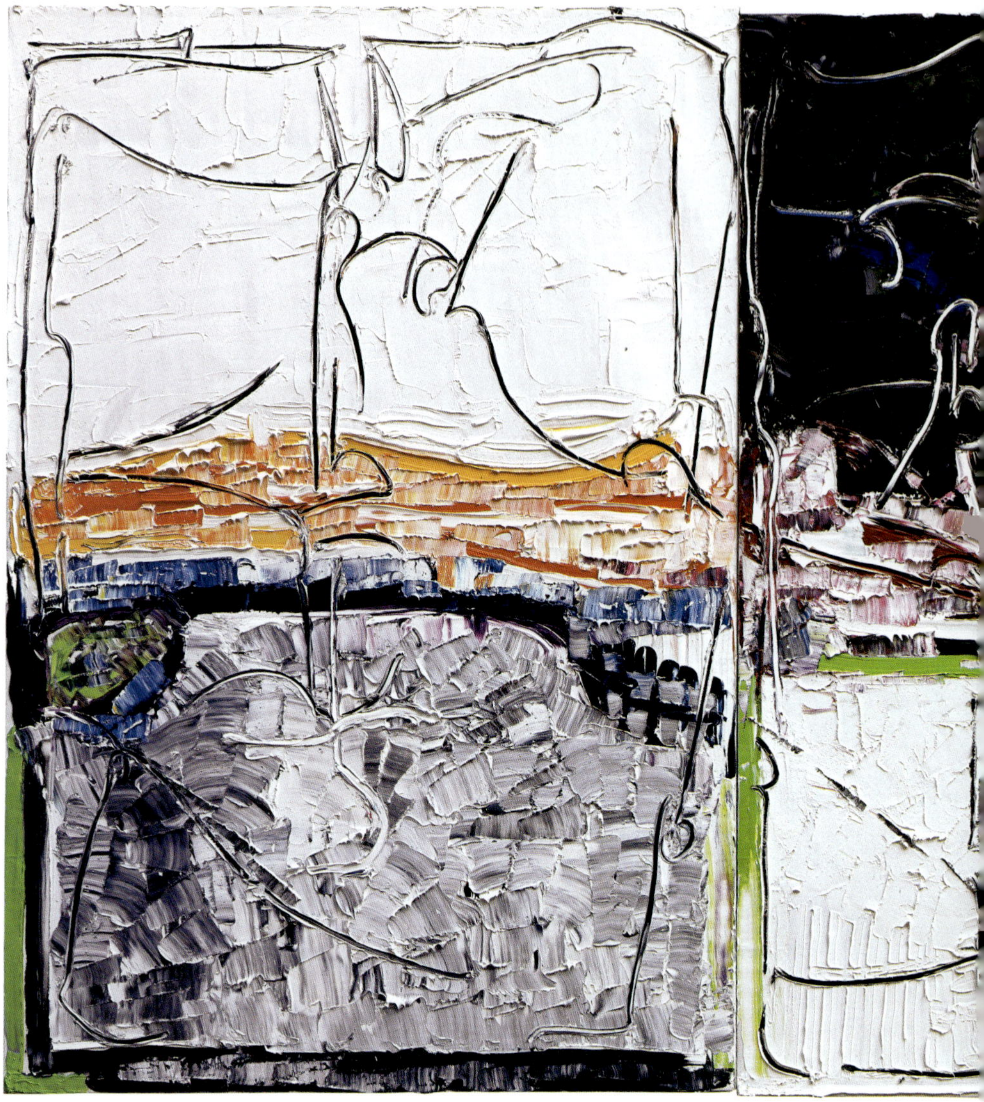

A/P ratios, the P/S ratio was above 2.0, too high for linseed oil alone and suggesting the presence of slower drying oils such as safflower or poppyseed.

It is important to note that the combination of cobalt blue or bone black pigment in a slow drying oil does not always lead to a weak paint layer with a high concentration of chain scission products. It is undoubtedly a combination of factors that has produced problematic blue, and to a lesser extent black, paints in Riopelle's works. His use of impasto, his thick paint layers, and the application of paint directly from the tube, in combination with his use of slow drying oils (such as safflower or poppyseed) and pigments with high oil absorption or slow drying, all work together to produce a paint film with a high concentration of chain scission products at the expense of cross-linking.

Exudates

Discolored accumulations of what looked like oil were observed on several paintings examined, most noticeable in large white areas in works from the 1960s and 1970s (fig. 30). Similar oil exudates were observed on works painted by Borduas in Paris in the late 1950s (Moffatt and Miller 1994). Samples of exudates—and, where possible,

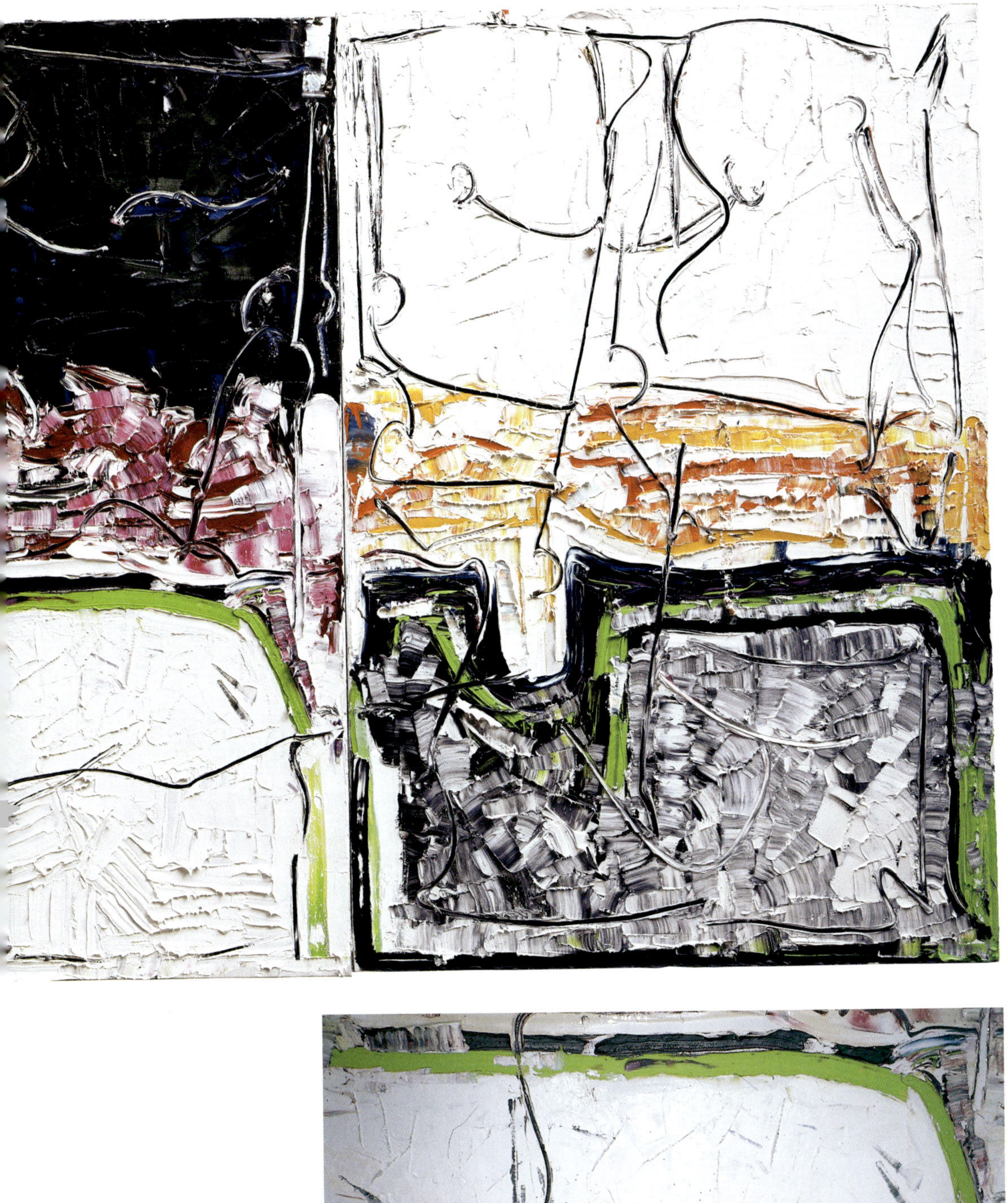

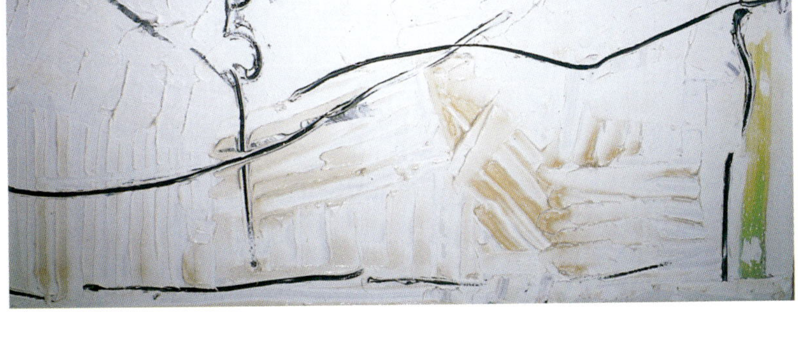

TABLE 4 Results of GC-MS analysis of oil exudates and associated white paint

Painting, date	P/S ratio* of exudate	A/P ratio† of exudate	P/S ratio of associated white paint	A/P ratio of associated white paint
Vent traversier, 1952	4.2	2.2	3.8	0.9
Poussière de soleil, 1954	1.6	2.4	—	—
Le Cirque, 1955	1.9	3.2	2.0	0.5
Point de rencontre, 1964	7.7	7.0	—	—
Mitchikanabikong, 1975	6.7	7.9	2.4	2.5

* P/S ratio = palmitate to stearate ratio.
† A/P ratio = azelate to palmitate ratio.

the white paint associated with the exudate—were taken from five of Riopelle's paintings dating from 1952 to 1975. The results of the GC-MS analysis for the exudates and associated white paints are shown in table 4.

All five exudates had elevated A/P ratios (greater than 2.0), indicating a high concentration of scission products in the fraction of oil that has separated from the paint layer. This is not surprising, since these oil exudates remain soft and tacky. The A/P ratios of the associated white paints are all lower than that of the exudates. The mechanism by which the oil has appeared on the surface of the white paint is not clear. However, it is unlikely that azelaic acid has migrated preferentially to palmitic and stearic acids through the paint film, as this diacid is rarely found in surface efflorescence or metal carboxylate soaps (Higgitt, Spring, and Saunders 2003). The exudates were more likely caused by paint that was not well mixed, leading to a separation of portions of the oil medium prior to drying and subsequent differences in the degree of cross-linking.

The P/S ratios of the oil exudates over white paint in *Vent traversier* and *Le Cirque* were similar to those of the white paint beneath. This is perhaps expected, since previous studies[34] indicate that the P/S ratio of blooms and soap inclusions is generally the same as that of the bulk paint. However, the oil exudate from *Mitchikanabikong*, dated 1975, showed a very high P/S ratio that did not match that of the associated white paint. This suggests that the P/S ratio may not remain constant in the case of oil that has separated from an uncured or partially cured paint layer.

Conservation Issues Observed in Mixed-Media Paintings

As fewer mixed-media paintings were included in the study and as these paintings are considerably more recent, very few conservation issues were noted.

Riopelle's use of nontraditional materials in the 1980s, including aerosol and paint for crafts and fabrics, introduced a new set of conservation concerns for his work. Paints based on an acrylic-acrylonitrile-styrene copolymer—including the metallic gold and silver paints and nacreous paints—used on the aerosol-painted background to create raised borders, contours, and designs have developed drying cracks (fig. 31). On several works, the brass powder used in gold-colored paint was found to have tarnished.

Most pigments identified through analysis proved to be stable to light. The majority of the organic pigments have a lightfastness that ranges from very good to excellent (Society of Dyers 1982). Exceptions could be fluorescent pigments and naphthol red,

FIGURE 31
Detail of *L'Oie hélico* showing drying cracks in black paint used to create raised borders. The background is painted with pink- and silver-colored aerosol paints. Lines created with silver- and gold-colored acrylic paint are also shown.

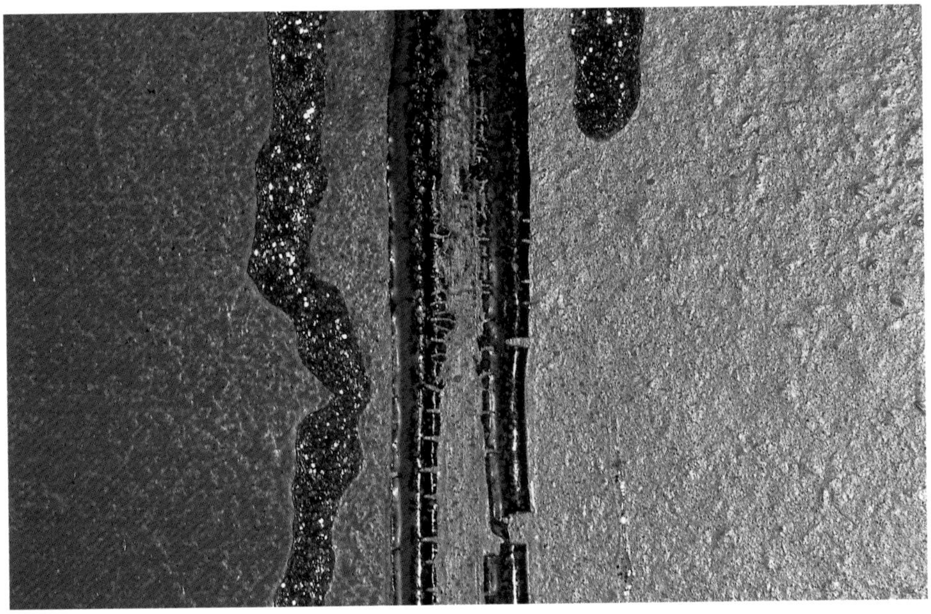

as there are several pigments in these categories with lightfastness ranging from fair to good (Voedisch 1973) and from fair to excellent (Society of Dyers 1982), respectively.

Aesthetic Characteristics of the Oil Paintings

The need to preserve volume, color, and range of gloss—the essential qualities that Riopelle obtained by selecting his oil paints with care—can present a considerable challenge to conservators who treat these paintings.

An issue of prime importance is varnishing. Conservators' observations and the authors' examination of a number of works confirmed that Riopelle did not varnish his paintings. Unfortunately, a number of Riopelle's works have been varnished, mostly by gallery owners who had paintings varnished after purchasing them and before putting them on the market. Oil paintings by Riopelle should never be varnished. Varnishing gives the entire painting a uniform gloss and neutralizes the contrast between matte and shiny areas that the artist sought. In addition, because the color and gloss of varnishes changes over time, applying a varnish could in the long term compromise the effect desired by the artist. Varnishing Riopelle's paintings would go against the artist's intent and is thus clearly prohibited by the conservators' code of ethics.[35]

Removing varnish from Riopelle's oil paintings cannot be done without risk to the works. For paintings on which there is considerable impasto, varnish removal can be very difficult, and the risk of damaging fragile areas, such as tips of impasto, is great. Furthermore, a number of conservators have noted that some colors used by Riopelle are very sensitive to the solvents required to remove varnish.

A second concern in the treatment of Riopelle's oil paintings is the juxtaposition of shiny and matte paint. Glossy, oil-rich paint, when applied in a thick layer, takes much longer than matte paint to dry. It may, in fact, take several decades to dry completely, depending on the composition and thickness of the paint layer. One often finds areas on Riopelle's works where paint is still malleable and others where paint can break off easily. When areas with soft, malleable paint undergo conservation procedures, such as

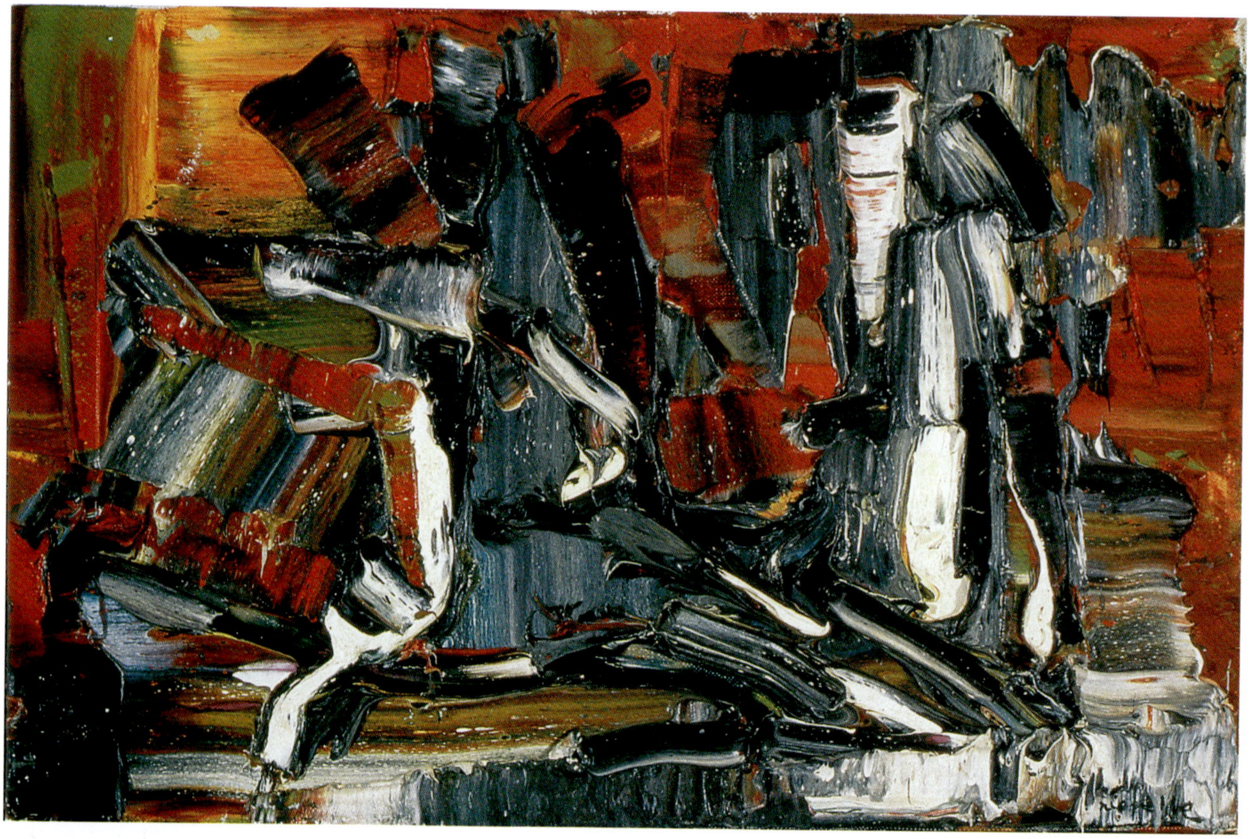

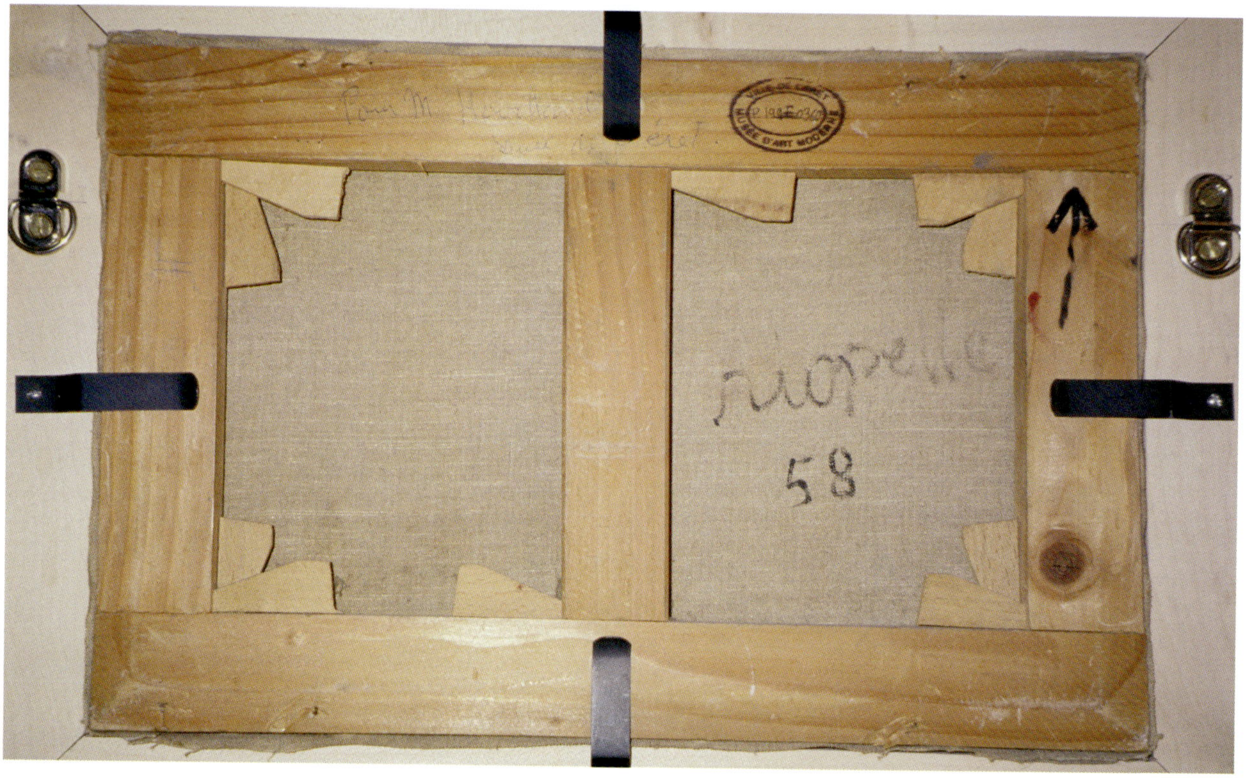
riopelle
58

FIGURE 32A-B
Jean Paul Riopelle, *Abstraction*, 1958, oil on canvas, 22 × 33 cm, Musée d'art moderne de Céret, inv. P1991-0360. Catalogue raisonné no. 1958.007H.1958. (a) Recto.

(B) Verso showing signature and date on the back of the canvas.

cleaning or consolidation, the use of pressure or heat should be minimized, since this could flatten or change the shape of the paint, thus affecting the three-dimensionality of the works, which is a very important component of Riopelle's œuvre.

Oil-rich, shiny paint that has not completely dried can remain sticky and attract dust and other airborne particulates. Excess oil can in some cases be observed on the surface of the paint, as was mentioned earlier. These oil exudates, which contain no pigment, can remain sticky for much longer than the paint beneath, since the presence of pigment makes paint dry more quickly. Oil exudates yellow with time, thus affecting the aesthetic qualities of works. It may be necessary to remove the oil film or, at least, to make it thinner in order to make the yellowing less noticeable.

At the opposite end of the spectrum from shiny paints rich in oil, one also finds matte paints in Riopelle's works. These paints can crack or become powdery with time. When paintings with matte paints are consolidated, the matte appearance must be preserved. Paintings that show such fragile, cracked paint, as well as those that show the lifting paint described earlier, must be manipulated and transported with great care to avoid further damage and possible paint loss.

Of course the primary concern must be with the front of Riopelle's paintings, but the back of the canvas should not be overlooked. Signatures or inscriptions on the back of the canvas or stretcher should be carefully documented before any conservation treatment that would cover or remove them (such as lining or replacement of the stretcher). Riopelle often signed and sometimes dated his works on the back of the canvas. Figure 32, for example, shows both the recto and the verso of *Abstraction*, dated 1958, a painting that Riopelle donated to Musée d'art moderne de Céret in France. While Riopelle's signature appears on the recto of the painting, the date is indicated on the verso only. In addition, there is an inscription in pencil on the stretcher (not legible in the figure) that reads, "Pour M. Havilland Musé de Céret." Although it was probably not written by Riopelle himself, this inscription provides a clue that the painting was intended for Frank Burty Haviland, one of the two founders of the Musée d'art moderne de Céret in 1950 (Lebovici 1999, 63). Based on important donations by Matisse and Picasso, the collection was developed essentially through gifts from artists, including Riopelle, who donated *Abstraction* in 1959.[36]

Conclusion

Riopelle's remarkable independence of spirit resulted in a body of work that, although rooted in Automatism, Surrealism, Abstraction, and the great modern pictorial tradition, developed according to the artist's own logic. The fact that Riopelle was not particularly interested in belonging to movements or schools may partly explain why his contribution and influence have not yet been fully recognized. This book explores the materials used in Riopelle's paintings, but there are many other aspects of his œuvre that deserve to be studied from an art historical point of view.

Throughout his career Riopelle demonstrated a profound and sincere interest in all the material aspects of his art. While he relied on experienced manufacturers to provide him with high-quality materials, he often used them in nonconventional ways. And whether he used traditional oil paint or more eclectic materials, Riopelle was always an innovative artist. With a relatively limited palette he created spectacular oil compositions. Later, he renewed his art by using nontraditional materials and mixed media. This move to new media is reflected in the color tones of Riopelle's paintings, with the mixed-media works often presenting bright, bold colors. Indeed, the pigments found in Riopelle's mixed-media paintings are quite different from those identified in his oil paintings. Most of the pigments in the oil paintings are inorganic; the majority of pigments in the mixed-media works are organic. The study of particular aspects of Riopelle's technique shows that modern paintings, whether created with oil paints or nontraditional materials, pose special conservation challenges.

Initially, Riopelle used oil paint, generously applied with a palette knife. His use of impasto, his thick paint layers, and the technique of applying paint directly from the tube, in combination with his use of slow drying oils and particular pigments, have led to conservation problems such as the lifting and flaking of paint as well as problematic drying of paint. One key aspect of his oil painting technique was the juxtaposition of matte and lustrous paint. The use of oil-rich, glossy paint, however, has led to the formation of oil exudates at the surface of certain works. These exudates are particularly disfiguring in some cases, ranging from dark yellow to almost brown. Riopelle did not varnish his paintings, and his works should not be varnished by others. Unfortunately, varnish was found in a number of samples from the oil paintings. In several paintings, varnishing has neutralized the effect Riopelle intended in his original juxtaposition of matte and glossy paint.

In the 1980s Riopelle experimented with nontraditional materials, such as aerosol paint and paint for crafts and fabrics. Fluorescent, metallic, and nacreous paints, as well as more traditional acrylics, work together to create paintings with a wide range of optical and textural effects. It is probably too early to say how these nontraditional materials will age, although some signs of deterioration, such as the tarnishing of the brass powder in gold-colored aerosol paint and cracking of paint based on a particular acrylic copolymer, have been noted already.

The scientific study of Riopelle's painting materials and techniques has provided information that will help answer questions related to the conservation of his works and the authenticity of paintings attributed to him. Even more important is that it has increased our understanding of the creative process of this major figure in Canadian painting.

OVERLEAF
Detail of Jean Paul Riopelle, *Rencontre*, 1956.

Photo: © Reinisches Bildarchiv Köln

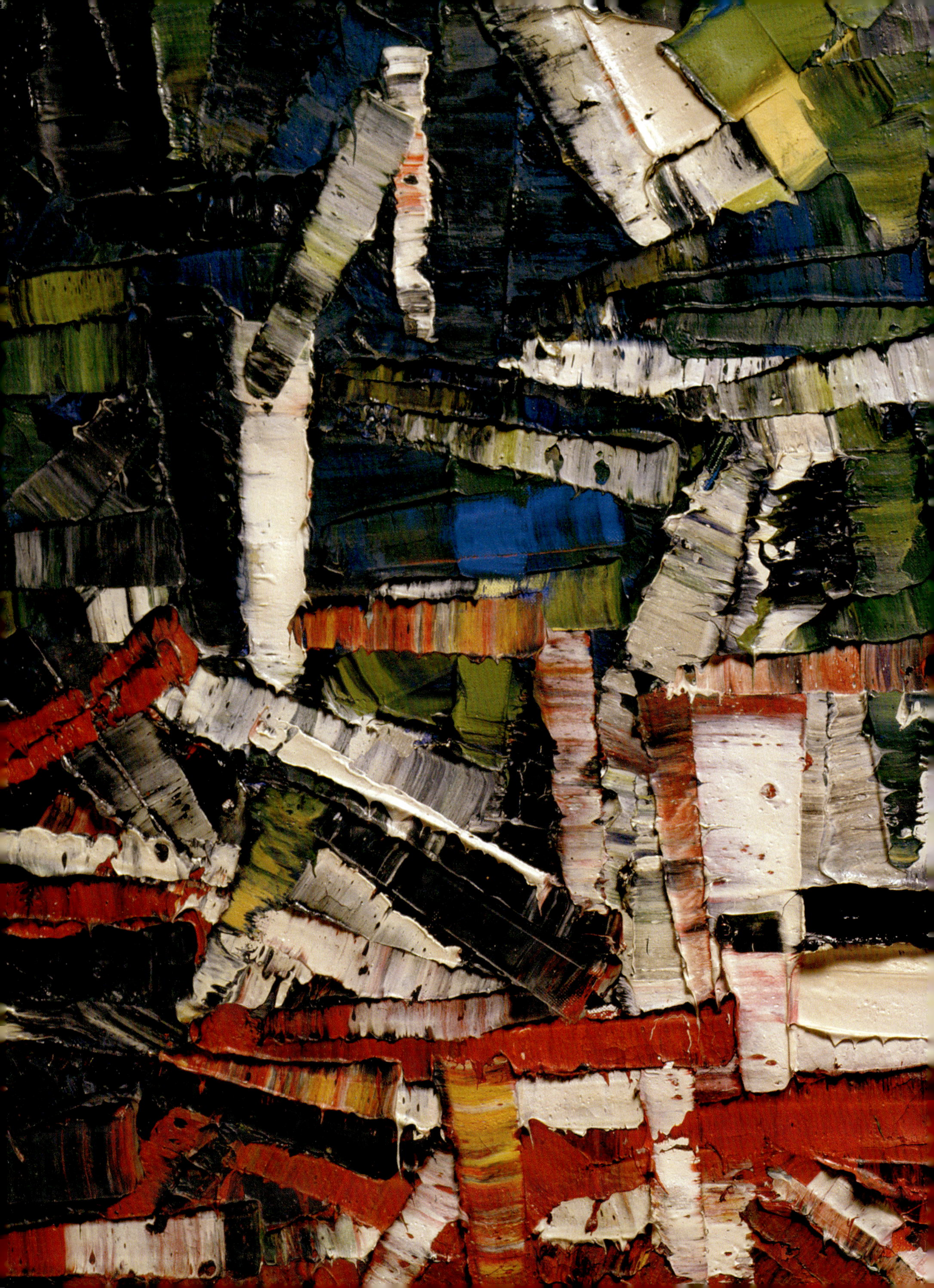

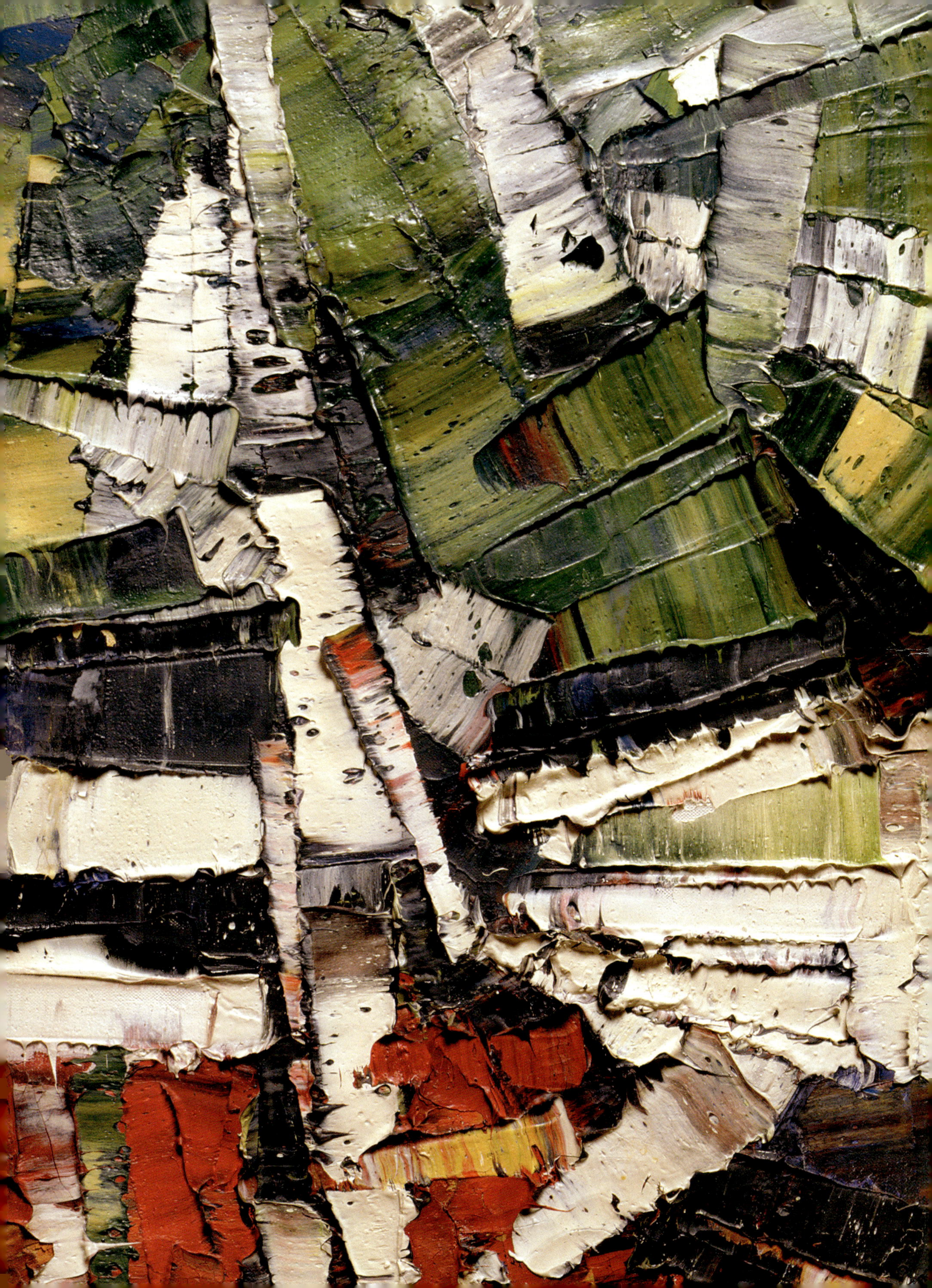

Appendix I

Analytical Methods

The elemental composition of particulate samples and individual layers in cross-sections was determined using a Hitachi S-3500N scanning electron microscope integrated with a light-element, lithium-drifted silicon, energy-dispersive X-ray detector, an Oxford Instruments Inca Wave 500 four-crystal spectrometer, and an Oxford Instruments Inca Energy+ X-ray microanalysis system.

Samples of ground and paint were analyzed by X-ray powder diffraction using a Rigaku RTP 300 RC generator equipped with a rotating anode and a cobalt target; the patterns were measured for 3600 to 7200 seconds using a microdiffractometer with a 100-μm collimator.

Samples were examined using FTIR spectroscopy. Most samples were analyzed using a Bomem Michelson MB100 spectrometer equipped with a 2-mm-diameter, wide-band mercury cadmium telluride (MCT) detector and a microbeam compartment that provides a 1.13-mm beam size at the focal point. Spectra were acquired from 4000 to 400 cm^{-1}. Many of the samples were also analyzed using a Bomem Michelson MB120 spectrometer interfaced to a Spectra-Tech IR-Plan Research microscope. The IR-Plan Research microscope is equipped with a 0.25-mm-diameter, narrow-band MCT detector with a cut-off at 660 cm^{-1}. Some samples were also analyzed using a Bruker Hyperion 2000 microscope equipped with a wide-band MCT detector with a cut-off at 430 cm^{-1} that was interfaced to a Tensor 27 spectrometer. Samples were masked with either fixed 100-μm circular apertures or variable rectangular apertures. Small samples, microsolvent extracts of multicomponent samples, specific areas of inhomogeneous samples and individual layers in multilayer samples were analyzed using this method. In all cases, the samples were mounted on a Spectra-Tech low-pressure diamond anvil cell composed of two Type IIA diamonds with a 0.6-mm-diameter free working area. Spectra were obtained at a resolution of 4 cm^{-1} by co-adding 150 to 200 interferograms.

Most samples of paint were prepared as pigment dispersions in Cargille Carmount mounting medium of refractive index of 1.66 (R.P. Cargille Laboratories, Inc.). The dispersions were observed with a Leica DMRX polarizing light microscope. Canvas samples were also examined by PLM.

GC-MS analysis was carried out on samples of ground, paint, and exudates. The GC-MS was run in both scan mode and selective ion monitoring (SIM) mode for each sample. Based on availability, samples of approximately 20 μg in size were placed in 2-mL GC vials, to which 20 μL of Meth Prep II (m-(trifluoromethyl)-phenyltrimethylammonium hydroxide (TMTFTH) 0.2 N in methanol) and 20 μL of toluene were added. In the cases of samples that were noticeably smaller, 15 μL of Meth Prep II and toluene were added to increase the concentration of the compounds of interest. The vials were then heated for 1 h at 65°C and centrifuged, and the extract was pipetted into a micro-vial insert and reinserted in the GC vial. Splitless injections of 1 μL were made, and the GC oven was programmed from 50°C to 200°C at a rate of 10°C/min and then from 200°C to 300°C at a rate of 6°C/min and a final hold time of 15 min; total run time was 46.67 min.

An Agilent 6890 gas chromatograph (GC) with 7683 auto injector interfaced to a 5973 mass selective (MS) detector was used with an HP-5MS capillary column ((5% phenyl)-methylpolysiloxane; 30.0 m, 0.25 mm internal diameter, 0.25 μm film thickness). The GC inlet temperature was set at 250°C and the MS interface at 280°C. Ultra-high purity helium carrier gas was used with a constant flow of 1.0 mL/min. The MS was run in scan mode from 50 to 600 amu, with the source and quadrupole temperatures set at 230°C and 150°C respectively. In SIM mode, three ions for each compound of interest were monitored at dwell times of 20 ms each during their elution window. The monitored ions used in the peak ratio calculations were as follows: dimethyl suberate (Su) (m/z 129.10, 138.10, and 171.10); dimethyl azelate (A) (m/z 152.10, 185.10, 143.10); internal standard (ISTD) hexadecane (m/z 57.00, 71.00, 226.2); methyl palmitate (P) (m/z 74.00, 227.20, 270.20); and methyl stearate (S) (m/z 74.00, 255.20, 298.20). The MS was operated in the electron impact (EI) positive ion mode (70 eV). Data were processed using Agilent Chemstation software (v.D.02.00.275).

Compound ratios were calculated using peak area integration. One scan and two SIM chromatograms were collected for each sample, and the peak ratios (P/S, A/P, Su/A, and P/ISTD) were calculated from both modes. Comparison of the calculations from the scan and SIM chromatograms normally indicated identical peak ratios.

Appendix II

Analysis of Riopelle's Oil Painting Medium

The drying oil medium in Riopelle's works was analyzed in 191 paint samples (from thirty-three paintings) and thirty-three ground samples (from twenty-three paintings, including two multicanvas paintings) using both FTIR spectroscopy and GC-MS. As well, samples of oil paint from a selection of paint tubes found in Riopelle's studio in France after his death were analyzed by the same methods.[37] The fatty acid profile of the drying oils in the samples, including the P/S ratio and the A/P ratio, was characterized using GC-MS (see Appendix I for details).

The Distribution of Drying Oil Types in Riopelle's Oil Paints: P/S Ratio

There was no indication from the analyses that Riopelle mixed his oil paint with other paint media. This was expected, based on his working method of applying paint directly from the tube with a palette knife. In a few samples from paintings, small amounts of beeswax, dammar, or a *Pinaceae* resin (such as pine or larch resin, for example) were identified in the paint by GC-MS along with the drying oil. These materials were undoubtedly added by the paint manufacturer; many of the paint tubes from Riopelle's studio that were analyzed contained beeswax, and some also contained dammar or *Pinaceae* resin.[38] None of Riopelle's oil paints or grounds showed evidence of having been heat pre-polymerized (i.e., stand oil).[39]

The P/S ratio remains relatively constant as drying oils age (Mills and White 1994, 171–72), and in certain circumstances this ratio can be used to determine the type of drying oil that has been used. However, as explained in more detail later, P/S ratios must be interpreted with care, especially in the case of modern paintings. A number of different drying oils have been employed in twentieth-century artists' oil paints, the most common being linseed, poppyseed, and safflower (Learner 2004, 20–21; Burnstock et al. 2006, 178–79). Safflower oil and poppyseed oil show less yellowing than linseed oil, so are often used in light-colored paints, such as whites and light blues, but have the disadvantage of drying more slowly than linseed oil. Typically, the P/S ratio of linseed oil is between 1.0 and 2.0 (with an average value of about 1.7), that of safflower oil is between 2.0 and 3.0 (with reported averages between 2.2 and 2.7), and that of poppyseed oil is between 3.0 and 8.0 (with an average value of about 3.3) (Colombini, Modugno, and Ribechini 2009, 199; Learner 2004, 21; Mills and White 1994, 33, 171–72; Gunstone, Harwood, and Padley 1986, 85). Analysis of standard drying oils at CCI using the procedure described in Appendix I produced P/S ratios in these ranges.

Figure 33 shows a histogram of the measured P/S ratios for the thirty-three ground samples from Riopelle's paintings that were analyzed by GC-MS. The histogram shows a distribution of values with a single maximum at a P/S ratio in the 1.5–1.7 range, a typical average value for linseed oil. It is clear from these data that the majority of the ground samples contain linseed oil as the binding medium. However, seven of the samples showed P/S ratios between 2.0 and 3.0, somewhat too high for linseed oil alone. In these cases, it is possible that another oil or oil mixture is present. The paintings with ground layers exhibiting P/S ratios above 2.0 did not come from a specific period in Riopelle's œuvre.

Figure 34 is a histogram of the measured P/S ratios for the 191 paint samples that were analyzed by GC-MS. Typical average values for linseed, safflower, and poppyseed oil are marked on the histogram for comparison with the data.

The histogram for the paint samples is more complex than that of the grounds. Theoretically, if only pure linseed, safflower, and poppyseed oils were used by Riopelle, one would expect to see an overlapping trimodal distribution of the data, with maxima at the average values of the three types of oil (similar to the bimodal distribution of walnut and linseed oil determined by Mills and White in historical paints from the fifteenth to the eighteenth century; Mills and White 1994, 172). This is not the case, however. The data show a range of P/S ratios, primarily between 1.0 and 5.0, with some outlying values. There is a maximum between about 1.5 and 2.0, with no other clear maxima in the distribution. The maximum in the histogram is close to the average value of linseed oil, indicating that many of Riopelle's paints contain linseed oil. The presence of samples with P/S ratios well above 3.0 indicates that some of his paints also contain poppyseed oil. Many samples gave intermediate P/S ratios, which are difficult to interpret.

Analysis of fresh paint from the interior of one of the Lefebvre-Foinet paint tubes from Riopelle's studio showed that Riopelle did use paint with a safflower oil medium. In fresh oil paint, for which the drying process has not yet begun, the full fatty acid profile (both saturated and unsaturated fatty acids) can be measured. Identification of the type of oil is based on the combination of the P/S ratio and the relative amounts of the unsaturated fatty acids oleic (C18:1), linoleic (C18:2), and linolenic (C18:3) acid. The medium of the paint from the Lefebvre-Foinet paint tube was identified as safflower oil based on the P/S ratio of 2.1 combined with the high proportion of linoleic acid and the absence of linolenic acid (Mills and White 1994, 33). By inference, then, it can be assumed that some of the intermediate P/S ratios in the samples from Riopelle's paintings also correspond to safflower oil. However, as figure 33 indicates, the data do not show the P/S ratios clustered in the ranges for specific oils, so care must be taken in the interpretation of individual P/S ratios, even if they fall into the typical range of a given oil. A number of factors may affect the P/S ratio in modern paints such as those used by Riopelle:

FIGURE 33

Histogram of the measured P/S ratios for the thirty-three ground samples from Riopelle's paintings analyzed by GC-MS.

FIGURE 34

Histogram of the measured P/S ratios for the 191 paint samples analyzed by GC-MS. Typical average values for linseed, safflower, and poppyseed oil are marked on the histogram for comparison.

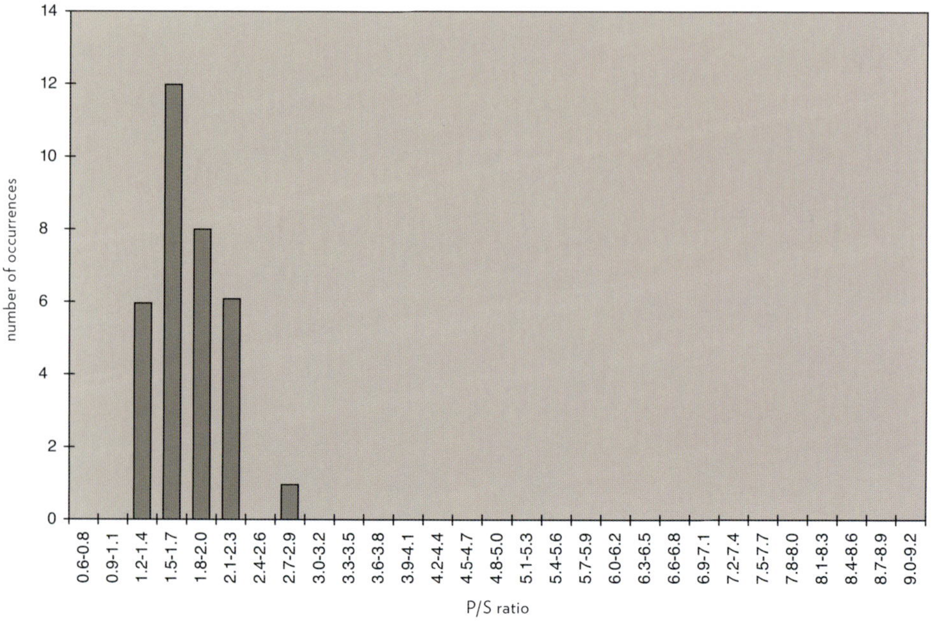

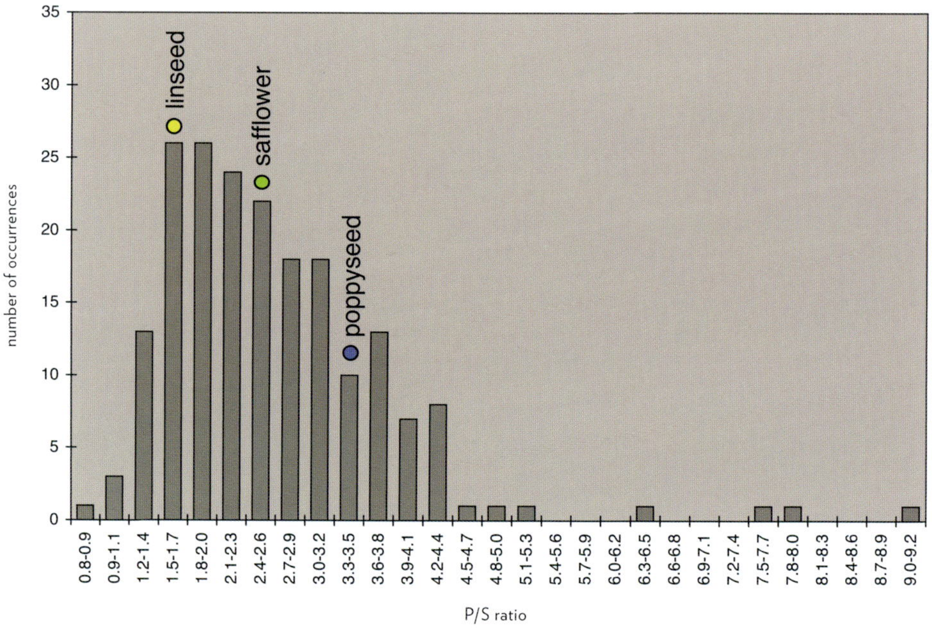

- Modern oil paints employ a wider range of oils and oil mixtures as compared with traditional oil paints (Learner 2004, 21). Some oils or oil mixtures have similar P/S ratios, and the oil type cannot be distinguished in a dried oil film. For example, a mixture of linseed and safflower oils (a mixture reported as being used by Winsor and Newton; Learner 2004, 21) would give a P/S ratio intermediate between that of the two pure oils. The intermediate P/S ratio produced from a mixture of linseed and poppyseed oil could be the same as that obtained for pure safflower oil.

- Some modern paints exhibit problematic drying, surface exudates, and soap formation. These issues are often related to the migration or reaction of free fatty acids within the paint film. In certain circumstances, palmitic and stearic acid migrate or react at different rates (see, e.g., Schilling, Carson, and Khanjan 1999, 244–45); this could lead to a variable P/S ratio over the surface or through the thickness of a paint film. A number of Riopelle's paints show problematic drying and surface exudates; preferential migration of palmitic acid could explain the outlying points in figure 33 with very high P/S ratios (as well as the high P/S ratio in two of the exudates described in chapter 5).

- Modern oil paints often contain driers and stabilizers based on metal fatty acid salts or waxes. Depending on their concentration, the presence of these compounds may change the measured P/S ratio of the paint. For example, a high proportion of aluminum stearate (a common additive in modern oil paints; Mayer 1991, 138–39; Learner 2004, 21–22; Tumosa 2001) would lower the measured P/S ratio.

FIGURE 35A–F
Measured A/P ratios for the 191 samples from Riopelle's oil paintings grouped by color (plotted as scatter plots from lowest to highest value).

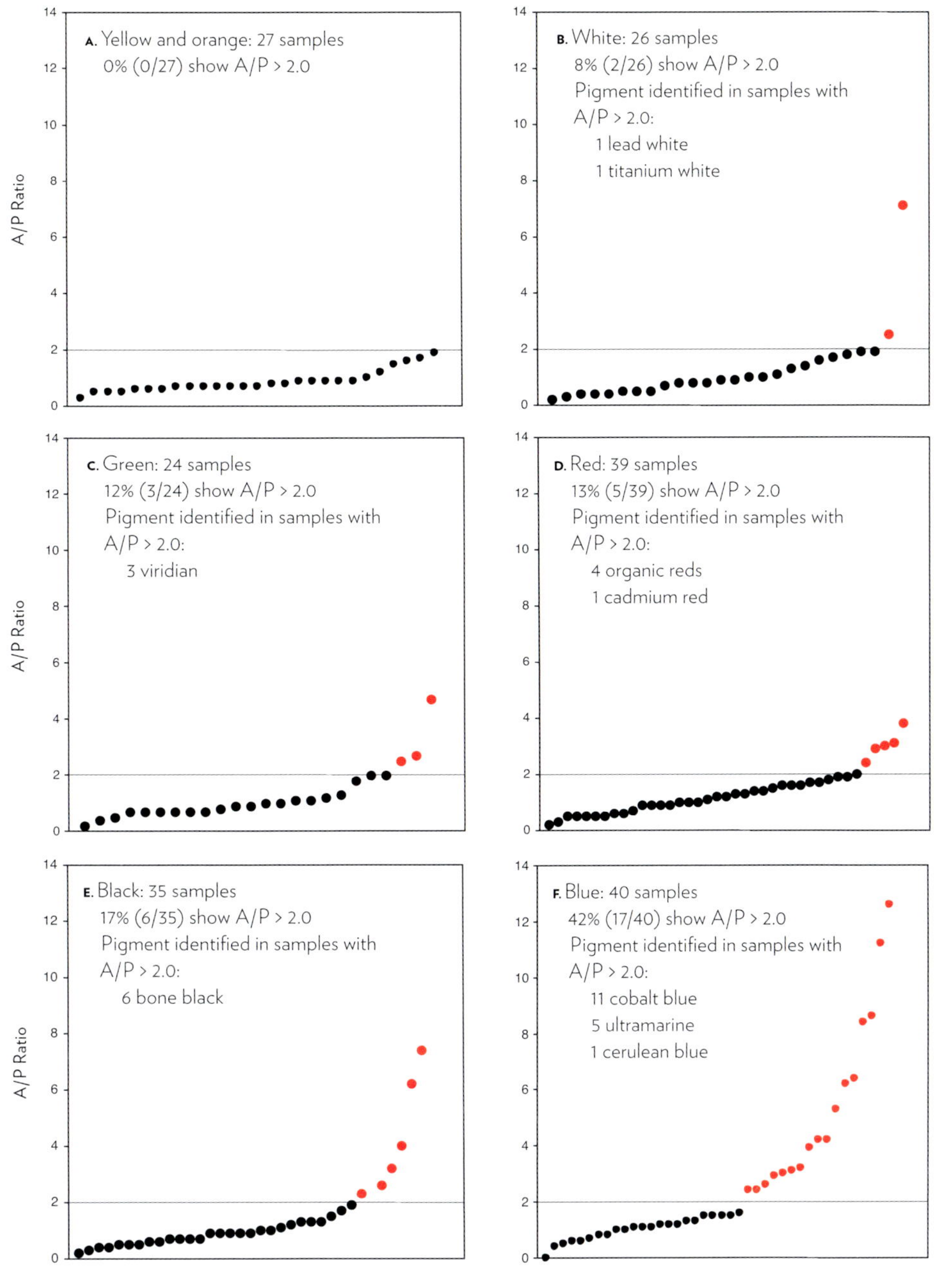

The Drying and Aging of Riopelle's Oil Paints

A/P Ratio

While the P/S ratio is used to determine the type of oil used, the A/P ratio can provide information about the drying and aging of the oil. Azelaic acid is the principal diacid that is formed during chain scission reactions that occur during both the curing and the aging of drying oil films (Mills and White 1994, 39–40; Colombini, Modugno, and Ribechini 2009, 198). A high A/P ratio indicates a high degree of chain scission at the expense of cross-linking; one can hypothesize, then, that paint exhibiting a high A/P ratio would produce a weaker, less cross-linked paint film (Miller and Moffatt 1994).

Figure 35 illustrates the measured A/P ratios for the 191 samples from Riopelle's oil paintings grouped by color. For each color group, the A/P ratios were plotted as scatter plots from the lowest to the highest value measured. The data were separated by color, since previous studies have shown that the type of pigment can have an effect on the A/P ratio, most likely by changing the amount of chain scission that occurs during drying or aging (Schilling and Khanjan 1996, 222–23). An examination of figure 35 shows that for most colors the measured A/P ratios are almost all below 2.0. However, the blue paints, and to a lesser extent the black paints, show a significant proportion of samples with higher A/P ratios. Over 40 percent of the blue paint samples studied had an A/P ratio greater than 2.0.

In paint samples that show an elevated A/P ratio by GC-MS, there is very often enough carboxylic acid present in the oil to be visible in the

FIGURE 36

FTIR spectrum of a cobalt blue paint with an elevated A/P ratio. The spectrum shows the ester band at 1734 cm^{-1} and carboxylic acid band at 1713 cm^{-1}.

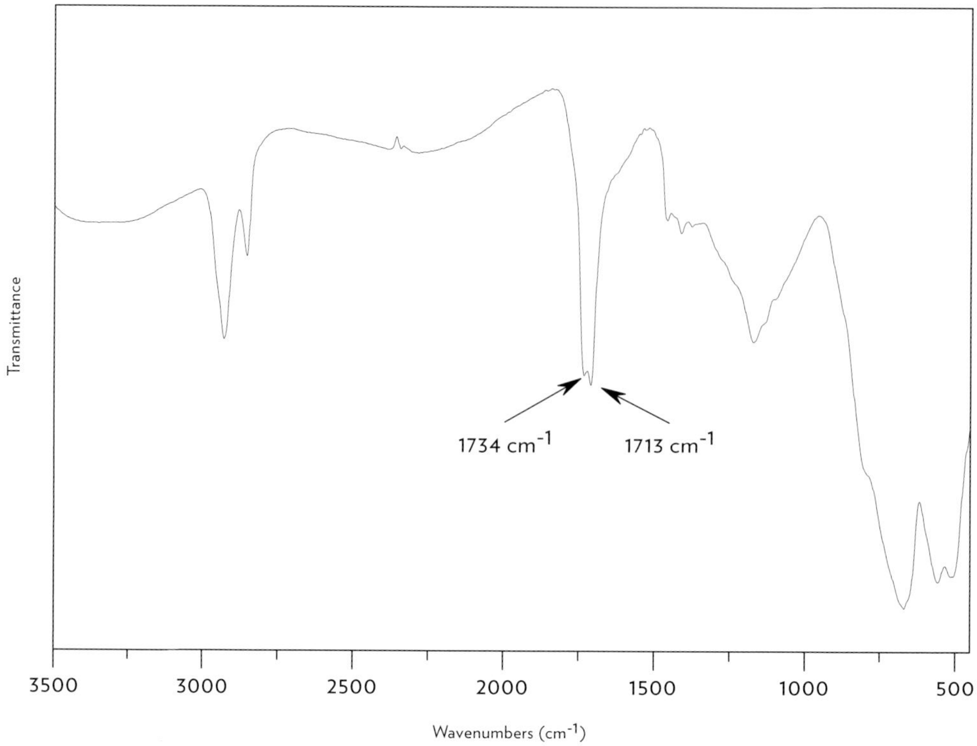

TABLE 5 Characteristics of the forty blue paint samples from Riopelle's oil paintings analyzed by GC-MS

Distribution of blue samples by pigment	Number (and percentage) of samples with elevated A/P ratio*	Number (and percentage) of samples with P/S ratio† > 2.0 among samples with elevated A/P ratio
13 cobalt blue	11 of 13 samples (85%)	10 of 11 samples (91%)
15 ultramarine blue	5 of 15 samples (33%)	4 of 5 samples (80%)
12 other blues‡	1 of 12 samples (8%)	1 of 1 samples (100%)

* A/P ratio = azelate to palmitate ratio. An elevated A/P ratio is defined as A/P > 2.0.
† P/S ratio = palmitate to stearate ratio.
‡ 6 cerulean blue; 4 Prussian blue; 1 PB 15; 1 unidentified blue.

FTIR spectrum without resolution enhancement or deconvolution. A split carbonyl band was observed in many of the infrared spectra with elevated A/P ratios. The spectra show the typical ester band for drying oils at 1730–1740 cm^{-1} and, in addition, a second band at 1710–1715 cm^{-1}, corresponding to carboxylic acids (van der Weerd, van Loon, and Boon 2005, 9). An example is shown in figure 36.

It is instructive to combine the results shown in figure 35 with visual observations of Riopelle's oil paintings. As explained in chapter 5, a number of Riopelle's works in oil showed blue paint with a problematic appearance (cracked, soft, or in some cases showing both powdery matte paint and shiny soft paint in the same area). Black paint was also sometimes cracked or soft, although less often than blue. The other colors were typically in good condition. These observations, combined with the data shown in figure 35, suggest a correlation between the condition of the paint and the A/P ratio. The reason that Riopelle's blue paints, and to a lesser extent his black paints, are problematic appears to be related to excessive chain scission during the drying or aging of the oil in these particular colors.

While bone black was the only black pigment that Riopelle used in his oil paintings, several different blue pigments were employed. Table 5 shows a more detailed breakdown of the data for the forty blue paint samples that were examined. Cobalt blue is the pigment most often present in blue paints with an elevated A/P ratio. In most cases, the cobalt blue was not mixed with other pigments or fillers. Of the thirteen cobalt blue paints examined, eleven showed an A/P ratio greater than 2.0. Although cobalt blue is considered to be an average drying pigment, it has a very high oil absorption index (Roy 2007, 157). This means that paints made with cobalt blue require a high concentration

FIGURE 37
FTIR spectra of different types of zinc soaps found in Riopelle's oil paints. (A) Zinc soaps with sharp bands matching zinc palmitate and/or stearate. (B) Zinc soaps with sharp bands matching zinc oleate. (C) Zinc soaps with broad bands, most likely due to different types of carboxylate groups within the cross-linked oil matrix.

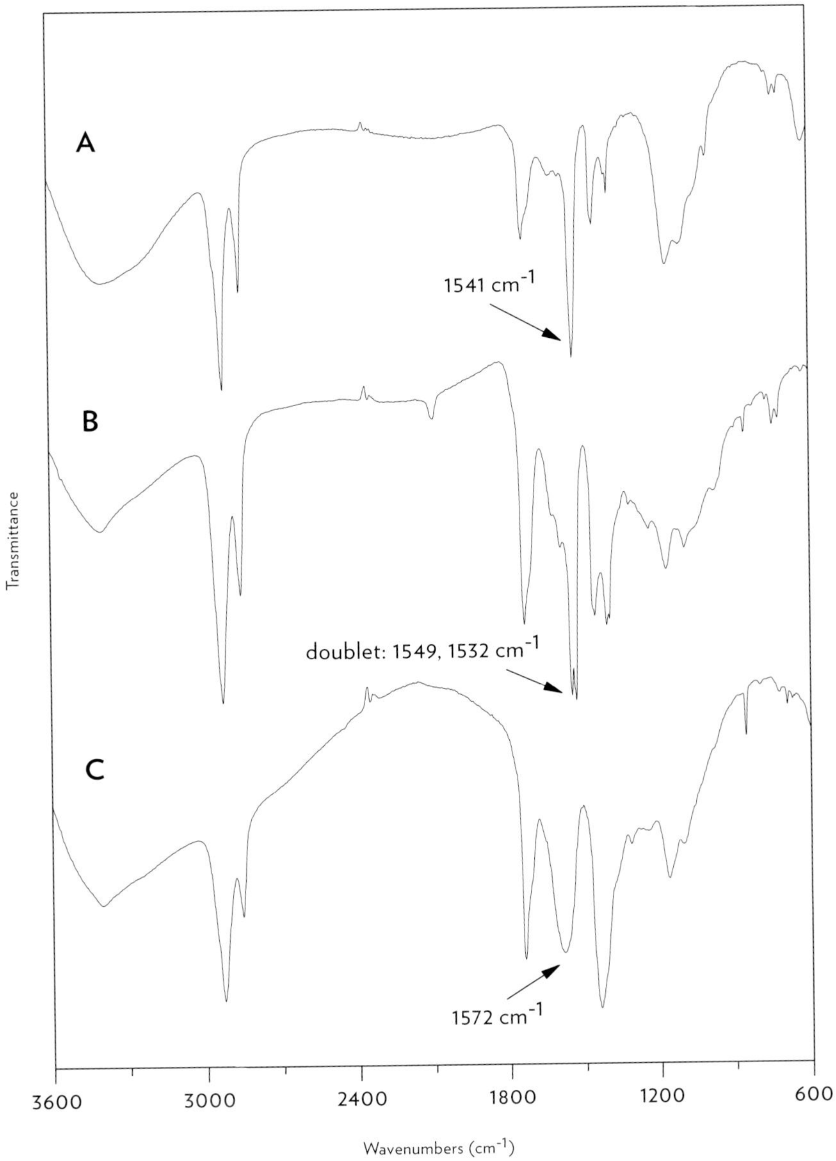

of oil medium, which in some situations could lead to problems with drying. Cracking and lifting cobalt blue paint with an elevated A/P ratio has been previously noted in a 1954 oil painting by Nicholas de Stael (Helwig and Miller 1995). Bone black, found in a smaller number of paints with an elevated A/P ratio, is a very slow drying pigment and also has a high oil absorption index (Mayer 1991, 133, 136). Problematic bone black paints, some with elevated A/P ratios, have been identified in the works of Paul-Émile Borduas (Moffatt and Miller 1994), who, like Riopelle, used Lefebvre-Foinet paints.

The type of drying oil used would also be expected to affect the amount of chain scission products and thus the A/P ratio. As shown in table 5, in almost all the cobalt blue paint samples with elevated A/P ratios, the P/S ratio was above 2.0, too high to correspond to linseed oil alone and suggesting the presence of slower drying oils such as safflower or poppyseed. The two cobalt blue paints with normal A/P ratios showed a P/S ratio below 2.0, suggesting the use of linseed oil. Similarly, five of the six bone black samples with an elevated A/P ratio showed a P/S ratio above 2.0, compared with only nine of twenty-nine blacks with normal A/P ratios.

While these trends are clear, it is important to point out that the combination of cobalt blue or bone black pigment in a slow drying oil does not always lead to a weak paint layer with a high concentration of chain scission products. Two of the cobalt blue samples and many of the bone black samples from Riopelle's oil paintings had A/P ratios in the normal range. As well, several cobalt blue samples taken from dried paints on the exterior of tubes in Riopelle's studio showed normal A/P values. Conversely, some blue paints in Riopelle's works that exhibited a cracked or powdery appearance and had elevated A/P ratios were pigmented with blues other than cobalt blue (Corbeil, Helwig, and Poulin 2004). As noted in chapter 5, it is undoubtedly a combination of factors—including Riopelle's use of specific pigments and oils along with aspects of his technique (impasto, thick paint layers, and the application of paint directly from the tube)—that has produced problematic blue, and to a lesser extent black, paints in some of his works.

Identification of Soaps in Paint Layers Using FTIR Spectroscopy

FTIR spectroscopy showed that most of the grounds and many of the paint samples in Riopelle's oil paintings contained lead or zinc soaps. In the case of lead soaps, the asymmetric COO^- stretch was observed at approximately 1515 cm^{-1}, consistent with lead fatty acid salts such as lead palmitate or stearate (Robinet and Corbeil 2003). The infrared spectra of paints containing zinc soaps, however, were more complex and showed that several different types of zinc fatty acid salts were present.

In some cases the zinc soaps produced sharp bands in the infrared spectrum, in good agreement with reference spectra of zinc palmitate and/or stearate (Robinet and Corbeil 2003). In these paints, the zinc soaps could be either additives in the paint or reaction products between zinc white and the oil matrix. An example of a spectrum showing sharp bands attributable to zinc palmitate and/or stearate is shown in the upper trace (A) of figure 37. In this spectrum, the strong, asymmetric COO^- stretch at 1541 cm^{-1} as well as several of the sharp, weak bands in the 1300–1150 cm^{-1} region due to CH_2 vibrations are present. Zinc soaps of this type have been identified in soap agglomerates in nineteenth- and twentieth-century paintings (Keune 2005, 145; van der Weerd et al. 2003, 410).

A few samples from Riopelle's oil paints showed sharp, well-resolved bands that match a reference spectrum of zinc oleate (Robinet and Corbeil 2003), a zinc soap that has not to our knowledge been previously reported in oil paints. Although prone to gradual oxidation, the mono-unsaturated oleic acid (C18:1) is often present in aged drying oils (Colombini, Modugno, and Ribechini 2009, 199; Mills and White 1994, 170–71). An FTIR spectrum of a sample from one of Riopelle's oil paintings showing bands consistent with zinc oleate is given in the middle trace (B) of figure 37. In this spectrum, the asymmetric COO^- stretch consists of a strong doublet at 1549 and 1532 cm^{-1} and a weaker absorption at 1593 cm^{-1}.

In most cases the zinc soaps in Riopelle's paints did not show well-resolved peaks but instead produced very broad absorptions centered at about 1570 and 1440 cm^{-1}. The band at 1570 cm^{-1}, corresponding to the asymmetric COO^- stretch, is at higher wavenumber and much broader than that of zinc palmitate or stearate. Similarly, the band at 1440 cm^{-1} is at higher wavenumber than the corresponding COO^- symmetric stretch and overlaps with the CH_2 bending vibrations in the 1470–1450 cm^{-1} region. These broad bands may be due to zinc ions coordinating with different types of carboxylate groups within the cross-linked oil matrix. Zinc soaps with these broadened, shifted bands have been observed in a study of naturally aged oil paint films that contain no added metal fatty acid salts (van der Weerd, van Loon, and Boon 2005, 18–19), providing evidence that the soaps are formed during the drying and aging of the oil. An example of an FTIR spectrum of this type of zinc soap is shown in the lower trace (C) of figure 37.

Notes

Introduction

1 The artist's first name is not hyphenated in this book, following the practice established in the catalogue raisonné; it reflects how Riopelle himself wrote his name (Riopelle 2010c). Quotations, however, follow the practice of the sources from which they are drawn.

2 Several fakes have been uncovered through CCI's scientific examination, and the authors are aware of at least one other fake being put on the market in Germany.

Chapter 1

3 Biographical information in this chapter is drawn from de Billy 1996, Bernier 1997, and Jolicoeur et al. 2002, which was largely based on documents from the Riopelle archives (Riopelle 2009a). A list of Riopelle's exhibitions is provided in Riopelle 2009a.

4 On his birth certificate, Léopold Riopelle's first name is written as Léon Paul. He is generally referred to, however, as Léopold (Riopelle 2011).

5 Riopelle mentioned that he started lessons when he was five years old (Letarte and Feaver 1982). Most sources state that formal lessons started with Bisson in 1936, when he was thirteen (de Billy 1996; Bernier 1997; Jolicoeur et al. 2002).

6 Michel Ragon, quoted in Nasgaard 2009, 7.

7 The signatories were (in alphabetical order) Madeleine Arbour, Marcel Barbeau, Paul-Émile Borduas, Bruno Cormier, Claude Gauvreau, Pierre Gauvreau, Muriel Guilbault, Marcelle Ferron, Fernand Leduc, Thérèse Renaud-Leduc, Jean-Paul Mousseau, Maurice Perron, Louise Renaud, Françoise Riopelle, Jean Paul Riopelle, and Françoise Sullivan.

8 Riopelle had in fact known Pierre Loeb for only five years, not the "seven or eight years" he mentioned when he told this story; they first met in 1947.

9 The same text was later used as a preface to the catalogue of Riopelle's first American solo exhibition (Duthuit 1954).

10 In 1945, to celebrate the liberation of France, Baron Philippe de Rothschild had the idea of dedicating the 1945 vintage to the Year of Victory; he commissioned the young artist Philippe Julian to produce a design that incorporated the "V" sign. From 1946 onward, a contemporary artist was commissioned to create an original artwork for the label each year. Artists such as Braque, Dalí, Miró, Chagall, Picasso, and Warhol preceded Riopelle. Riopelle actually produced two designs, but since it was difficult to choose between them, both were used (http://www.bpdr.com/, accessed December 8, 2009).

11 The Government of Canada donated this painting to France in 1989, for the bicentennial of the French Revolution. The painting is now housed at the Paris Bastille Opera House.

12 http://www.fondation-maeght.com/, accessed December 2, 2009.

13 Although one source notes that "Riopelle often said that he had been practicing the art of sculpture 'forever'" (Daigneault 2009, 401), only three early clay sculptures are reproduced in the catalogue raisonné, all dated 1947–48 (Riopelle 1999, 422–24).

14 This treatment was applied to original works on paper as well as to prints, lithographs, and silkscreens that Riopelle reworked in ink or acrylic paint (a technique he had first begun to use in the late 1960s). Adhering works on paper to canvas was not Riopelle's doing, except for the large collages he created in 1967, which were attached to large canvases (491 cm × 243 cm) from the beginning (Riopelle 2010c).

15 Most sources say that Riopelle returned to Canada in 1989. Yseult Riopelle has specified, however, that he returned on the first day of 1990 (Riopelle 2010a).

Chapter 2

16 All quotations in this chapter from French-language sources have been translated by the author.

17 The art historian Serge Guilbaut conducted a close analysis of the more global issues in which Riopelle played a key role, notably through his influence on expatriates from 1955 to 1979 (Guilbaut 2002).

18 On this issue, see Guilbaut 1983.

19 As the master engraver René Le Moigne recalled, Riopelle brought him leaves, pieces of canvas, and other unusual materials for prints and asked him for things that no other artist had requested: "I was the first—perhaps others have done so since—who made prints of tree leaves, burlap or embossed materials without a raster, photo or anything else: a direct print on stone or zinc with an inking process, transfer process, etc." (Le Moigne 2004, 68).

Chapter 3

20 Gagnon 1988, 42; English translation from Waldberg 1999, 46. (English translations of French-language sources quoted in chapters 3, 4, and 5 are by the authors unless otherwise noted, as here.)

21 Brunet-Weinmann, however, mentions that Riopelle was using paint directly out of the container in which he had ground his paints during the time when he made his own (Brunet-Weinmann 1999, 132).

22 Early colorman stamps observed on the paintings of the Australian

artist Rupert Bunny indicate that Paul Foinet once had a store on rue Notre-Dame-des-Champs and that its name evolved from "Paul Foinet" to "Paul Foinet Fils" and then to "Lucien Lefebvre-Foinet" (Ives and Nottage 2009).

23 The colorman stamp was normally applied on canvases mounted on stretchers, although it was sometimes applied on unmounted canvases at the request of clients (Établissements Lefebvre-Foinet 2002).

24 Synthetic plumbonacrite was identified by X-ray diffraction in the grounds of paintings by Soulages (Corbeil and Helwig 2008) and Borduas. In the latter case, X-ray diffraction was not performed as part of the published study on Borduas (Moffatt and Miller 1994); recent analysis using this technique confirmed its presence.

25 According to Viau (2002, 48), Riopelle stopped using brushes in 1949.

26 While this photograph was most likely staged (Riopelle 2010b), several people from Riopelle's entourage made reference to piles of tubes in his studio.

27 Viau (2002, 48) reported that Riopelle used sticks to throw strings of paint.

28 Colour Index numbers are provided for the organic pigments, since they are often referred to with these numbers.

29 This cobalt violet is a cobalt magnesium arsenate, $Mg_2Co(AsO_4)$, a composition that was reported and discussed in Corbeil, Charland, and Moffatt 2002.

Chapter 4

30 Not all these doors were used at Île-aux-Oies. *Hommage à Duchamp*, which became *Hommage à Maurice Richard*, was painted in Sainte-Marguerite-du-Lac-Masson (Riopelle 2010b).

31 Catalogue notice by François-Marc Gagnon in Montreal Museum of Fine Arts 2002, 106.

Chapter 5

32 See, for example, Moffatt and Miller 1994; O'Malley and Moffatt 2001; Robinet and Corbeil 2003, 23–24.

33 The intense band at 1515 cm^{-1}, as well as the series of sharp, weak bands in the 1300–1150 cm^{-1} region, due to CH_2 vibrations, allowed the soap to be identified as primarily lead palmitate (Robinet and Corbeil 2003).

34 Summarized in Higgitt, Spring, and Saunders 2003.

35 See, for example, the clause titled "Respect for the Integrity of the Cultural Property" in the *Code of Ethics and Guidance for Practice* of the Canadian Association for Conservation of Cultural Property and Canadian Association of Professional Conservators (2000, 4): "When conserving a cultural property, the conservation professional shall respect the integrity of the cultural property by endeavouring to preserve its material composition and culturally significant qualities through minimal intervention. The original intention, usage, history and evidence of provenance of the property must be respected. This respect for the integrity of the cultural property shall be based upon the study of the cultural property and on consultations with the owner and, when applicable, the originator. When relevant, other authorities or documentary sources should be consulted."

36 The date of the gift is specified in the painting's dossier.

Appendix II

37 While specific additives or materials in the tubes were helpful in interpreting the results from the paint samples, the overall results from the tubes are limited in number and from a specific time period and thus do not represent Riopelle's œuvre as a whole. For this reason, the results for the tubes are given only in cases where they aid interpretation of the paint samples.

38 Waxes and natural resins have both been used as additives by paint manufacturers (Mayer 1991, 185–86). Dammar in particular is included in an oil painting medium recommended by Mayer as "the best for general, all-around purposes" (Mayer 1991, 251).

39 Drying oil with a high concentration of linolenic acid, such as linseed oil, can be determined to be heat pre-polymerized when trace peaks of cyclic C18 fatty acids (methyl 9-(2-propylphenyl)nonanoate, and isomers thereof having a molecular weight of 290 amu) that are formed by cyclization of linolenic acid upon heating are present in the scan mode chromatograms (Perkins and Iwaoka 1973; van den Berg, van den Berg, and Boon 2002). Additionally, in cases where these trace compounds are present, the A/P ratio is lower than 1.0, and the suberate to azelate (Su/A) ratio is slightly elevated at 0.4–0.5 (Pitthard et al. 2005).

Bibliography

Adam, Édouard (of Adam Montparnasse). 2010. Personal communication with Marie-Claude Corbeil, March 14.

Aquin, Stéphane. 2002. Riopelle's critical migration. In *Riopelle* (*Connaissance des arts*, special issue no. 179), 33–40. Montreal: Montreal Museum of Fine Arts.

Arbour, Madeleine, and Jeanette Tardif. 1972. *Riopelle: Ficelles et autres jeux*. VHS. Montreal: Société Radio-Canada.

Béguin, André. 1981. *Dictionnaire technique de la peinture*, vol. 1. Paris: André Béguin.

Bellet, Harry. 1991. The chasse-galerie. In *Jean-Paul Riopelle*, 25–42. Montreal: Montreal Museum of Fine Arts.

Bernier, Robert. 1997. *Jean-Paul Riopelle: Des visions d'Amérique*. Montreal: Éditions de l'Homme.

Bertrand, Jean-Guy. 1991. "Je m'en irai avec les oies." *Journal de Montréal* (September 21), 30.

Bond Rafferty, Jean. 2009. In Paris, a colorful collection comes to auction. *ARTINFO.com* (November 25). http://www.artinfo.com/news/story/33304/in-paris-a-colorful-collection-comes-to-auction/.

Borduas, Paul-Émile. 1985. *Refus Global/Total Refusal*. Trans. Ray Ellenwood. Toronto: Exile Editions.

Breton, André, Élisa Breton, and Benjamin Péret. 1949. Aparté entre Élisa, André Breton et Benjamin Péret. Preface to *Riopelle à la Dragonne, Galerie Nina Dausset, 23 mars–23 avril 1949*, unpaginated. Paris: Galerie Nina Dausset.

Brunet-Weinmann, Monique. 1999. Birth of a signature. In *Jean Paul Riopelle, catalogue raisonné*, vol. 1, *1939–1953*, ed. Yseult Riopelle, 99–142. Montreal: Hibou Éditeurs.

———. 2009. The matter, the hand. In *Jean Paul Riopelle, catalogue raisonné*, vol. 3, *1960–1965*, ed. Yseult Riopelle and Tanguy Riopelle, 21–40. Montreal: Hibou Éditeurs.

Burnstock, Aviva, Klaas Jan van den Berg, Suzan de Groot, and Louise Wijnberg. 2006. An investigation of water-sensitive oil paints in twentieth-century paintings. In *Modern Paints Uncovered*, ed. Thomas J. S. Learner, Patricia Smithen, Jan W. Krueger, and Michael R. Schilling, 177–88. Los Angeles: Getty Conservation Institute.

Cage, Otis. 1954. The reflective eye. *Art Digest* 28 (7): 4, 25–26.

Canadian Association for Conservation and Canadian Association of Professional Conservators. 2000. *Code of Ethics and Guidance for Practice of the Canadian Association for Conservation of Cultural Property and Canadian Association of Professional Conservators*. 3rd ed. Ottawa: Canadian Association for Conservation and Canadian Association of Professional Conservators.

Challan-Belval, Marie-Noël. 1991. *Materials, Techniques and Conservation Problems Related to the Paintings of Paul-Émile Borduas from 1955 to 1960*. Master of Art Conservation thesis, Queen's University.

Colombini, Maria Perla, Francesca Modugno, and Erika Ribechini. 2009. GC/MS in the characterization of lipids. In *Organic Mass Spectrometry in Art and Archeology*, ed. Maria Perla Colombini and Francesca Modugno, 191–214. West Sussex, UK: John Wiley and Sons.

Corbeil, Marie-Claude. 2004. Considerations regarding Riopelle's pictorial technique and conservation. In *Jean Paul Riopelle, catalogue raisonné*, vol. 2, *1954–1959*, ed. Yseult Riopelle, 19–25. Montreal: Hibou Éditeurs; Moudon: Acatos.

Corbeil, Marie-Claude, Jean-Pierre Charland, and Elizabeth A. Moffatt. 2002. The characterization of cobalt violet pigments. *Studies in Conservation* 47 (4): 237–56.

Corbeil, Marie-Claude, and Kate Helwig. 2008. *Analyse d'échantillons de tableaux de Pierre Soulages*. Ottawa: Canadian Conservation Institute. Unpublished report.

Corbeil, Marie-Claude, Kate Helwig, and Jennifer Poulin. 2004. Analysis of the painted œuvre of Jean-Paul Riopelle: From oil to mixed media. In *Modern Art, New Museums: Contributions to the Bilbao Congress, 13–17 September 2004*, ed. Ashok Roy and Perry Smith, 170–73. London: International Institute for Conservation of Historic and Artistic Works.

Corbeil, Marie-Claude, Kate Helwig, and Jennifer Poulin. 2006. Une étude scientifique de la technique picturale de Jean-Paul Riopelle. *Techne* 24, 47–52.

Corbeil, Marie-Claude, and P. Jane Sirois. 2007. A note on a modern lead white, also known as "synthetic plumbonacrite." *Studies in Conservation* 52 (4): 281–88.

Daigneault, Gilles. 2009. The pleasure of sculpture, prolegomena to an analysis of Riopelle's three-dimensional work. In *Jean Paul Riopelle, catalogue raisonné*, vol. 3, *1960–1965*, ed. Yseult Riopelle and Tanguy Riopelle, 399–411. Montréal: Hibou Éditeurs.

de Billy, Hélène. 1996. *Riopelle*. Montreal: Éditions Art Global.

Duthuit, George. 1952. A painter of awakening: Jean-Paul Riopelle. *Canadian Art* 10 (1), 24–27.

———. 1954. A painter of awakening: Jean-Paul Riopelle. Trans. Samuel Beckett. In *Riopelle: First American Exhibition. Catalogue of an Exhibition Held at the Pierre Matisse Gallery, January 5–23, 1954*, unpaginated. New York: Pierre Matisse Gallery.

Ellenwood, Ray. 1992. *Eggregore: A History of the Montréal Automatist Movement*. Toronto: Exile Editions.

Érouart, Gilbert. 1995. *Riopelle in Conversation: With an Interview by Fernand Séguin*. Trans. Donald Winkler. Concord, Ont.: House of Anansi Press.

Établissements Lefebvre-Foinet. 2002. Personal communication with Marie-Claude Corbeil, April 3.

Franz, Klaus-Dieter, Hartmut Härtner, Ralf Emmert, and Katsuhisa Nitta. 1993. Nacreous and interference pigments. In *Industrial Inorganic Pigments*, ed. Gunter Buxbaum, 207–21. Weinheim: VCH.

Gagnon, Daniel. 1988. *Riopelle grandeur nature*. Cap Saint-Ignace, Quebec: Éditions Fides.

Gagnon, François-Marc. 1998. *Chronique du mouvement automatiste québécois: 1941–1954*. Montreal: Lanctôt Éditeur.

———. 2000. L'Hommage à Rosa Luxemburg. In *Riopelle au Musée du Québec: L'Hommage à Rosa Luxemburg*. Quebec: Musée du Québec.

Greenberg, Clement. 1953. Symposium: Is the French avant-garde overrated? *Art Digest* 27 (20): 12, 27.

Guilbaut, Serge. 1983. *How New York Stole the Idea of Modern Art: Abstract Expressionism, Freedom, and the Cold War*. Trans. Arthur Goldhammer. Chicago: University of Chicago Press.

———. 2002. From earth to sky with Riopelle. In *Riopelle* (*Connaissance des arts*, special issue no. 179), 17–24. Montreal: Montreal Museum of Fine Arts.

Gunstone, F. D., John L. Harwood, and F. B. Padley. 1986. *The Lipid Handbook*. London: Chapman and Hall.

Helou-de la Grandière, Pauline, Anne-Solenn Le Hô, and François Mirambet. 2008. Delaminating paint films at the end of 1950s: A case study on Pierre Soulages. In *Preparation for Paintings: The Artist's Choice and Its Consequence*, ed. Joyce H. Townsend, Tiarna Doherty, Gunnar Heydenreich, and Jacqueline Ridge, 156–62. London: Archetype Publications.

Helwig, Kate, and David Miller. 1995. Examination of the painting *Port de Sicile* by Nicholas de Stael. Ottawa: Canadian Conservation Institute. Unpublished report.

Higgitt, Catherine, Marika Spring, and David Saunders. 2003. Pigment-medium interaction of oil paint films containing red lead or lead-tin yellow. *National Gallery Technical Bulletin* 24: 75–95.

Hunter, Sam. 1954. Riopelle: Conquest of space. *Art Digest* 28 (8): 12.

Ives, Simon, and Andrea Nottage. 2009. La vie boheme: Rupert Bunny and the colourmen of Montparnasse. *Look, Art Gallery of New South Wales* (September): 32–34.

Johnson, Ken. 2005. Jean-Paul Riopelle. *New York Times,* April 22, E33.

Jolicoeur, Michel, Sophie Malouin, Benoît Giguère, and Robert Bernier. 2002. *Salut, Riopelle!* DVD. Montreal: TRAM Design Multimédia.

Keune, Katrien. 2005. Binding medium, pigments and metal soaps characterized and localized in paint cross-sections. PhD diss., FOM Institute for Atomic and Molecular Physics.

Learner, Thomas J. S. 2004. *Analysis of Modern Paints*. Los Angeles: Getty Conservation Institute.

Lebovici, Elisabeth. 1999. 1940–1965: Le temps de la mémoire, vers la création du musée. In *Guide: Musée d'art moderne de Céret*, ed. Joséphine Matamoros, 57–73. Céret: Musée d'art moderne de Céret.

Le Moigne, René. 2004. In *Riopelle vu par*...Texts and interviews gathered by Michel Waldberg. Paris: Centre culturel canadien.

Letarte, Pierre, and Marianne Feaver. 1982. *Riopelle*. VHS. Ottawa: National Film Board of Canada.

Livingston, Jane. 2002. *The Paintings of Joan Mitchell*. New York: Whitney Museum.

Marchbank, Joseph. 1974. Aerosol coatings. In *Technology of Paints, Varnishes and Lacquers*, ed. Charles R. Martens, 607–12. Huntington, NY: Robert E. Krieger Publishing.

Mayer, Ralph. 1991. *The Artist's Handbook of Materials and Techniques*. 5th ed. New York: Viking Penguin.

Michaud, Yves. 1994. La rêverie de la nature. In *Jean-Paul Riopelle: Les années soixante*, unpaginated. Paris: Didier Imbert Fine Art.

———. Organicity and cosmic flow. In *Jean Paul Riopelle, catalogue raisonné,* vol. 3, *1960–1965*, ed. Yseult Riopelle and Tanguy Riopelle, 45–64. Montreal: Hibou Éditeurs.

Mills, John S., and Raymond White. 1994. *The Organic Chemistry of Museum Objects*. 2nd ed. Oxford: Butterworth-Heinemann.

Moffatt, Elizabeth A., and David Miller. 1994. Spectroscopic and chromatographic analysis of selected paintings from the Parisian period of Paul-Émile Borduas. *Journal of the Canadian Association for Conservation* 19: 3–13.

Montreal Museum of Fine Arts. 1991. *Jean-Paul Riopelle*. Montreal: Montreal Museum of Fine Arts.

———. 2002. *Riopelle* (*Connaissance des arts*, special issue no. 179). Montreal: Montreal Museum of Fine Arts.

Moos, David. 2004. Pavane: Notes on the relationship between Jean Paul Riopelle and Joan Mitchell. In *Jean Paul Riopelle, catalogue raisonné*, vol. 2, *1954–1959*, ed. Yseult Riopelle, 125–26. Montreal: Hibou Éditeurs; Moudon; Acatos.

Nasgaard, Roald. 2009. The Automatiste revolution in painting. In *The Automatiste Revolution*. Vancouver/Toronto/Berkeley: D&M Publishers.

O'Malley, Michael, and Elizabeth A. Moffatt. 2001. Paul-Émile Borduas: The blacks and whites of the later paintings (1955–1960). In *Deterioration of Artists' Paints: Effects and Analysis (Joint Meeting of ICOM-CC Working Groups Paintings 1 & 2 and The Paintings Section, UKIC, British Museum, London, 10th and 11th September 2001: Extended Abstracts of Presentations)*, 84–87. London: United Kingdom Institute for Conservation of Historic and Artistic Works (UKIC) and International Council of Museums (ICOM).

Perkins, Edward G., and Wayne T. Iwaoka. 1973. Purification of cyclic fatty acid esters: A GC-MS study. *Journal of the American Oil Chemists' Society* 50: 44–49.

Pitthard, V., S. Stanek, M. Griesser, and T. Muxeneder. 2005. *Chromatographia* 62 (3–4): 175–82.

Riopelle, Yseult, ed. 1999. *Jean Paul Riopelle, catalogue raisonné,* vol. 1, *1939–1953*. Montreal: Hibou Éditeurs.

———, ed. 2005. *Jean Paul Riopelle, catalogue raisonné, tome des estampes*. Montreal: Hibou Éditeurs.

———. 2009a. *Le catalogue raisonné de Jean Paul Riopelle*. http://www.riopelle.ca/

———. 2009b. Personal communication with Marie-Claude Corbeil, January 7.

———. 2010a. Personal communication with Marie-Claude Corbeil, February 15.

———. 2010b. Personal communication with Marie-Claude Corbeil, February 24.

———. 2010c. Personal communication with Marie-Claude Corbeil, March 10.

———. 2010d. Personal communication with Marie-Claude Corbeil, August 1.

———. 2011. Personal communication with Marie-Claude Corbeil, January 27.

Riopelle, Yseult, and Tanguy Riopelle, ed. 2009. *Jean Paul Riopelle, catalogue raisonné,* vol. 3, *1960–1965*. Montreal: Hibou Éditeurs.

Robinet, Laurianne, and Marie-Claude Corbeil. 2003. The characterization of metal soaps. *Studies in Conservation* 48 (1): 23–40.

Roy, Ashok. 2007. Cobalt blue. In *Artists' Pigments: A Handbook of Their History and Characteristics*, vol. 4, ed. Barbara Berrie, 151–77. Washington, DC: National Gallery of Art.

Schilling, Michael R., David M. Carson, and Herant P. Khanjan. 1999. Gas chromatographic determination of the fatty acid and glycerol content of lipids. IV. Evaporation of fatty acids and the formation of ghost images by framed oil paintings. In *Preprints: ICOM Committee for Conservation, 12th Triennial Meeting, Lyon, August 29–September 3, 1999*, vol. 1, ed. Janet Bridgland, 242–347. London: James and James (Science Publishers).

Schilling, Michael R., and Herant P. Khanjan. 1996. Gas chromatographic determination of the fatty acid and glycerol content of lipids. I. The effects of pigments and aging on the composition of oil paints. In *Preprints: ICOM Committee for Conservation, 11th Triennial Meeting, Edinburgh, September 1–6, 1996*, vol. 1, ed. Janet Bridgland, 220–27. London: James and James (Science Publishers).

Schneider, Pierre. 1972. *Parler de corde*. Paris: Maeght Éditeurs.

Sheffy, Pearl. 1964. Jean-Paul Riopelle talks about art. *Globe Magazine* (May 9): 6–8.

Spalding, Jeffrey. 2002. Immaculate misconception. In *Riopelle* (*Connaissance des arts*, special issue no. 179), 9–15. Montreal: Montreal Museum of Fine Arts.

Society of Dyers and Colourists and American Association of Textile Chemists and Colorists. 1982. *Colour Index*. 3rd ed. Bradford, UK: Society of Dyers and Colourists and American Association of Textile Chemists and Colorists.

Tumosa, Charles S. 2001. A brief history of aluminum stearate in paint. *WAAC Newsetter* 23 (3): 10–11.

Vachon, Huguette. 2000. Ambiance. In *Riopelle: A Space of Freedom*, 93–98. Trois-Rivières: Éditions d'art Le Sabord.

Van den Berg, Jorrit Dirk Jan, Klaas Jan van den Berg, and Jaap J. Boon. 2002. Identification of non-crossed-linked compounds in methanolic extracts of cured and aged linseed oil–based paint films using gas chromatography-mass spectrometry. *Journal of Chromatography A* 950: 195–211.

Van der Weerd, Jaap, Muriel Geldof, Luuk Struik van der Loeff, Ron M. A. Heeren, and Jaap J. Boon. 2003. Zinc soap aggregate formation in "Falling Leaves (Les Alyscamps)" by Vincent van Gogh. *Kunsttechnologie Konservierung* 17 (2): 407–16.

Van der Weerd, Jaap, Annelies van Loon, and Jaap J. Boon. 2005. FTIR studies of the effects of pigments on the aging of oil. *Studies in Conservation* 50 (1): 3–22.

Viau, René. 1991. Deux Jean du pays: Une saison dans la vie de Riopelle. *La Presse*, September 21.

———. 2002. Jamais deux fois le même tableau. *Vie des arts* 187: 48–49.

Voedisch, Robert W. 1973. Luminescent pigments, organic. In *Pigment Handbook*, vol. 1, ed. Temple C. Patton, 891–903. New York: John Wiley & Sons.

Waldberg, Michel. 1999. Riopelle, the absolute gap. In *Jean Paul Riopelle, catalogue raisonné*, vol. 1, *1939–1953*, ed. Yseult Riopelle, 39–54. Montreal: Hibou Éditeurs.

———. 2004. Riopelle, ondoyant et divers. In *Riopelle vu par*... Texts and interviews gathered by Michel Waldberg. Paris: Centre culturel canadien.

Index

Acknowledgments

The authors wish to thank curators and conservators at the following institutions for granting permission to examine and sample paintings by Riopelle in their collections: Montreal Museum of Fine Arts, Musée d'art contemporain de Montréal, Musée d'art moderne de Céret (France), Musée national d'art moderne (France), and Musée national des beaux-arts du Québec.

The authors are especially grateful to the following conservators for sharing their knowledge of conservation issues in regard to Riopelle's paintings: Marion Barclay and Richard Gagnier, formerly at the National Gallery of Canada; Claude Belleau, Musée national des beaux-arts du Québec; Marie-Noël Challan-Belval, Musée d'art contemporain de Montréal; David Cueco, private conservator, Paris; Catherine O'Meara, Montreal Museum of Fine Arts; and Debra Daly-Hartin and Helen McKay, Canadian Conservation Institute.

The authors would also like to express their most sincere gratitude to Yseult Riopelle, without whose support and assistance the research project, and consequently this book, would not have been as complete. Her interest in our project and her generosity in sharing information with us were most appreciated and are gratefully acknowledged.

This project benefited from the technical support of two interns, Shawn Hopewell and Philip Cook. Shawn Hopewell's Young Canada Works in Science and Technology internship at the Canadian Conservation Institute was funded by the Department of Canadian Heritage.

Finally, the authors are grateful to Tom Learner, senior scientist at the Getty Conservation Institute, for his invitation to publish this book in the GCI's Artist's Materials series. We would like to thank Tim Whalen, director of the GCI, for his support throughout this project. The preparation and publication of this manuscript benefited from the skills of many talented individuals; we would particularly like to express our appreciation to Cynthia Godlewski, senior publications specialist at the CGI; the book's copy editor, Katya Rice; and, at Getty Publications, Tevvy Ball, who worked assiduously developing the manuscript; Jeffrey Cohen, who created the book's handsome design; and Suzanne Watson, who coordinated its production.

About the Authors

STÉPHANE AQUIN has worked as curator of contemporary art at the Montreal Museum of Fine Arts (MMFA) since 1998. He is responsible for the collection from 1945 to the present, including a large number of works by Jean Paul Riopelle. He coordinated the 1991 MMFA retrospective exhibition devoted to the artist and curated the 2002 MMFA retrospective exhibition as well as the traveling exhibition presented at the State Hermitage Museum (St. Petersburg) and the Musée Cantini (Marseille) in 2006.

MARIE-CLAUDE CORBEIL received a B.Sc. in chemistry from Université de Montréal. She then specialized in inorganic chemistry and crystallography and completed a master's program in 1984 and a Ph.D. program in 1987. She has worked at the Canadian Conservation Institute (CCI) as a conservation scientist since 1988, performing analyses of museum objects, especially paintings, and is currently manager of the Conservation Science Division at CCI.

KATE HELWIG received a B.Sc. in chemistry from the University of Toronto, a master's degree in physical chemistry from Stanford University, and a master's degree in art conservation from Queen's University. She has worked as a conservation scientist at CCI since 1992. She specializes in the analysis of art and archaeological objects using a diverse range of techniques, including Fourier transform infrared spectroscopy.

JENNIFER POULIN received a B.Sc. with honors in chemistry from Acadia University and a master's degree in analytical chemistry from Dalhousie University, specializing in gas chromatography. She has worked in the analysis of natural products since 1996 and began work at CCI in 2003 as a conservation scientist. She specializes in the analysis of organic components in art and archaeological objects using gas chromatography-mass spectrometry.

YSEULT RIOPELLE is the eldest daughter of the choreographer Françoise Lespérance and the painter Jean Paul Riopelle. In 1985 she undertook the catalogue raisonné of Riopelle's œuvre. In this capacity, she was mandated by Riopelle to conduct research to gather all information about his œuvre, to be the editor of the catalogue raisonné, and to authenticate his works. She has collaborated on many exhibitions of Riopelle's works in Europe and in Canada, as well as on books and documentaries about him.